Mechanical Design of Microresonators

Mechanical Design of Microresonators

Modeling and Applications

Nicolae Lobontiu

McGraw-Hill

New York Chicago San Francisco Lisbon London Madrid
Mexico City Milan New Delhi San Juan Seoul
Singapore Sydney Toronto

The *McGraw·Hill* Companies

Library of Congress Cataloging-in-Publication Data

Lobontiu, Nicolae.
   Mechanical design of microresonators: modeling and applications/
Nicolae Lobontiu.
      p.   cm.
   Includes bibliographical references and index.
   ISBN 0-07-145538-8
   1. Microelectromechanical systems—Design and construction.
   2. Electric resonators—Design and construction.   3. Vibrators—
Design and constuction.   4. Detectors—Design and construction.
I. Title.

   TK7875.L62   2006
   621.381—dc22

                                           2005051140

1 2 3 4 5 6 7 8 9 0   DOC/DOC   0 1 0 9 8 7 6 5

ISBN 0-07-145538-8

*The sponsoring editor for this book was Kenneth P. McCombs, the editing
supervisor was David E. Fogarty, and the production supervisor was
Pamela A. Pelton. It was set in Century Schoolbook by Digital Publishing
Solutions. The art director for the cover was Handel Low.*

Printed and bound by RR Donnelley.

 This book was printed on recycled, acid-free paper
containing a minimum of 50% recycled, de-inked fiber.

McGraw-Hill books are available at special quantity discounts to use as
premiums and sales promotions, or for use in corporate training
programs. For more information, please write to the Director of Special
Sales, McGraw-Hill Professional, Two Penn Plaza, New York, NY
10121-2298. Or contact your local bookstore.

With love to my wife Simona
and daughters Diana and Ioana

Contents

Preface

Mechanical microresonators are fundamental components in a host of MEMS applications covering the automotive sector (safety systems, stability and rollover, occupant detection, tire pressure monitoring, biometric sensors for comfort programs), the telecommunication industry [especially the radio-frequency (RF) domain with implementations such as switches, tunable capacitors and mechanical filters implemented in wavelength division multiplexing (WDM) and mobile communication, variable attenuators in cell phones, frequency reference, digital micromirror devices (DMD), laser tuning or radar systems], the bio/medicine domain [detection and tracking of various substances including hazardous and explosive ones at the femtogram level, magnetic resonance imaging (MRI), surgical instrumentation for corneal resurfacing or hair/tattoo removal], the material/surface characterization area [scanning probe microscopy (SPM) and atomic force microscopy (AFM), resonant strain gauges, residual stress measurements], and motion sensing (gyroscopes and other resonant accelerometers for navigational systems and platform stabilization). Applications of mechanical micro-resonators are also implemented in virtual reality, people-to-device communication (gloves, helmets and haptic systems for remote surgical intervention), optical beam scanners, laser printers, inertial mouse devices in computers, CD players, video cameras, fluid density and mass/pressure flow sensors, low acceleration (low-g) sensors, and light modulators.

Based on the resonant beam technology, mechanical microresonators are capable of high accuracy and sensitivity (order of magnitudes over conventional-technology counterparts), very good signal-to-noise (S/N) ratio, relatively large bandwidth, compatibility with the integrated-circuit (IC) technology, simplified digital interface, and miniaturization. The last feature in this enumeration is crucial, particularly in detecting minute amounts of substances (at the cell level in biodetection, for instance) where very small quantities of extraneous matter can be detected only by very small mechanical resonators which operate at frequencies in the gigahertz domain. The large spectrum of current and foreseeable micromechanical resonator applications has sparked a wide interest in advancing the practical and theoretical knowledge in this area. Inroads have been made at all component levels that are involved in developing high-performance mechanical resonator systems, including the fabrication, electronic, mechanical, and control subdomains. This book is dedicated to the mechanical modeling and design of microresonators. The book addresses the main methods and procedures which can be utilized in evaluating the behavior of mechanical microresonators by means of lumped- and distributed-parameter modeling. It also contains a database offering comprehensive characterization of mechanical resonator components and systems (including hinges, cantilevers and bridges)—many of them novel—in the frequency domain. It is hoped that professionals with various expertise levels and backgrounds, who are involved with the study, research, and development of mechanical microresonators will find this book useful. Many fully solved, real-life resonator examples accompany and complement the basic material. Although many of today's mechanical resonators are fabricated in the nanometer range, the prefix *micro* has been used in this book to keep the nomenclature unitary and short-form.

Chapter 1 introduces the main traits of modeling and designing mechanical microsystems which operate at resonance. Single- and multiple-degree-of-freedom systems are characterized in terms of their free and forced response. The damping in mechanical microsystems is discussed including loss mechanisms such as those produced by fluid-structure interaction and internal dissipation. Methods enabling us to formulate the dynamic equations of motion and to determine the resonant response, both exact and approximate, are also presented in Chapter 1, which concludes with notions of mechanical-electrical analogies, transfer functions, complex impedances, and micromechanical resonator filters.

Chapter 2 focuses on basic components that are the backbone of mechanical microresonators, such as line members, circular rings, thin plates, and membranes. Lumped-parameter modeling is presented together with the methods enabling derivation of stiffness properties (Castigilano's displacement theorem) and inertia fractions (Rayleigh's principle), which are usually combined to yield the relevant resonant frequencies. Basic microcantilever shapes such as constant cross-section, trapezoid and corner-filleted are fully defined in terms of their axial, torsional, and bending resonant frequencies. The distributed-parameter modeling approach targets the resonant characterization of line members under the action of axial loads, circular rings, thin plates and membranes.

Chapters 3 and 4 are dedicated to microhinges, microcantilevers, and microbridges. These compliant members can be utilized as either stand-alone resonator systems (such as in mass detection or switches) or components of more complex resonators (such as elastic suspensions). The lumped-parameter stiffness, inertia and corresponding resonant frequencies are derived for various configurations including paddle, filleted (circular and elliptic), notched, hollow, and multimorph (sandwiched). Generic formulations are also provided which facilitate modeling and designing of components with geometric profiles other than the ones presented in these chapters.

Chapter 5 studies resonant mechanical microsystems such as beam type, spring type, microgyroscopes, tuning forks, and microaccelerometers. Various models are proposed and compared, which characterize the dynamic response and performance of mechanical microresonators at different levels of accuracy. The main methods of transduction (actuation and/or sensing) which are implemented in microresonator applications such as electrostatic, electromagnetic, piezoelectric and piezomagnetic are also discussed in Chapter 5.

The final chapter, Chapter 6, focuses entirely on microcantilever and microbridge systems which are designed for mass detection. Static detection of extraneous substance attachment is treated here but the emphasis falls on resonant methods and devices enabling mass detection by means of resonant frequency shift monitoring.

The book contains quite a few novel designs and associated models, and although a lot of effort and time has been spent at making sure that the mathematical apparatus is correct, errors might have slipped in— I would appreciate signaling of such occurrences.

My thanks go to Dr. Rob Ilic of Cornell NanoScale Facility for allowing me to present pictures of his work on microresonators, for his enthusiastic and thorough review of the chapter on mass detection, and for the precious suggestions on the introduction to this chapter, which have been included almost *ad literam*.

NICOLAE LOBONTIU
Cluj-Napoca, Romania

1

Design at Resonance of Mechanical Microsystems

1.1 Introduction

This chapter is an introduction to the main aspects encountered in modeling and designing mechanical microresonators.

Aside from the technological reasons for realizing systems that integrate the mechanical structure and the associated silicon/semiconductor electronic circuitry, the drive toward smaller-scale, nano-domain mechanical resonators is motivated by the need for pushing the limits to the resonant frequencies in the gigahertz domain. It is known that the stiffness of a mechanical resonator varies with the inverse of the length (because the basic definition of stiffness is force divided by length):

$$k \sim \frac{1}{l} \tag{1.1}$$

and that the resonant frequency is proportional to the square root of the stiffness:

$$\omega_r \sim \sqrt{k} \tag{1.2}$$

As a consequence, increasing the resonant frequency of a mechanical device implies miniaturization, and therefore very high frequencies are achieved by very small resonator dimensions. In addition, as this chapter discusses, higher resonant frequencies (which are achieved

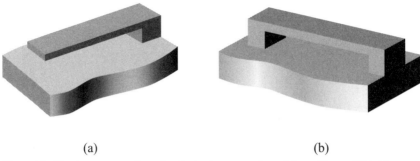

(a)   (b)

Figure 1.1   Single-component mechanical microresonators: (*a*) cantilever; (*b*) bridge.

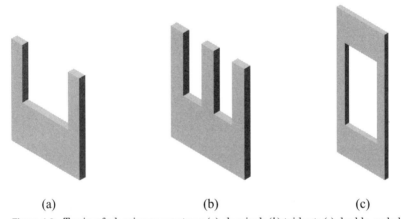

(a)   (b)   (c)

Figure 1.2   Tuning-fork microresonators: (*a*) classical; (*b*) trident; (*c*) double-ended.

with small-dimension resonators) also contribute to increasing the quality factor of a system, which is a measure of its resonant performance. Smaller is also better, as Chap. 6 will demonstrate, in detecting minute amounts of deposited substances as the capacity of capturing the effects of mass at the cell level is inversely proportional to the geometric dimensions of a mechanical resonator.

Constructively, the mechanical microresonators can be cantilevers, as sketched in Fig. 1.1*a*; bridges, as in Fig. 1.1*b*; tuning forks, as shown in Fig. 1.2*a, b,* and *c.* Or they can be of a more complex geometry, such as the lateral resonator design with folded-beam suspensions illustrated in Fig. 1.3. More details regarding these mechanical resonators, as well as more resonator structures, are presented in subsequent chapters of this book.

This chapter analyzes the main aspects of single- and multiple-degree-of-freedom mechanical microresonators by discussing the models that are utilized to characterize and design these devices.

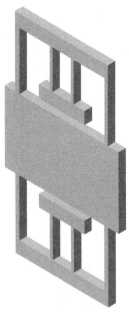

Figure 1.3 Lateral mechanical microresonator with folded-beam suspensions.

1.2   Single-Degree-of-Freedom Systems

Many mechanical microresonators can be modeled as single-degree-of-freedom systems. A microcantilever, for instance, such as the one illustrated in Fig. 1.4, may only vibrate in bending and therefore can be modeled as a single-degree-of-freedom member by means of lumped-parameter properties (as shown in subsequent chapters in this book), namely, by allocating mass and stiffness fractions at the free end about the single motion direction. The free response of a mechanical system determines the resonant frequency in either the presence or the absence of damping. The forced response reveals the behavior of an undamped or damped mechanical system under the action of a sinusoidal (most often) excitation. In mechanical resonators, the phenomenon of resonance is important, and in such situations the excitation frequency matches the natural (resonant) frequency of the system.

1.2.1   Free response

For a single-degree-of-freedom (single-DOF) system formed of a body of mass m and a spring of stiffness k, such as the one in Fig. 1.5, the dynamic equation of motion is

$$m\ddot{x} + kx = 0 \qquad (1.3)$$

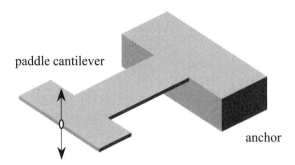

paddle cantilever

anchor

translatory motion

Figure 1.4 Microcantilever as a single-degree-of-freedom system.

Figure 1.5 Single-degree-of-freedom mass-spring system.

The solution to Eq. (1.3) is

$$x(t) = \frac{\dot{x}_0}{\omega_r} \sin(\omega_r t) + x_0 \cos(\omega_r t) \tag{1.4}$$

where the natural (or resonant) frequency is

$$\omega_r = \sqrt{\frac{k}{m}} \tag{1.5}$$

and the initial displacement and velocity conditions are

$$x(0) = x_0 \quad \left.\frac{dx}{dt}\right|_{t=0} = \dot{x}_0 \tag{1.6}$$

Similarly, the equation of motion of a single-DOF system formed of a mass and a dashpot (mass-damper combination with viscous damping), such as the one in Fig. 1.6, is

$$m\ddot{x} + c\dot{x} + kx = 0 \tag{1.7}$$

and the solution to this homogeneous equation can be expressed as

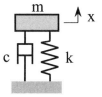

Figure 1.6 Single-degree-of-freedom mass-dashpot system.

$$x(t) = [x_0\cos(\omega_d t) + (\dot{x}_0 + \xi\omega x_0)/\omega_d \sin(\omega_d t)]e^{-\xi\omega t} \tag{1.8}$$

where

$$\omega_d = \sqrt{1-\xi^2}\,\omega_r \tag{1.9}$$

is the damped frequency of the system and the damping ratio ξ is defined as

$$\xi = c/c_c = c/(2\sqrt{mk}) = c/(2m\omega_r) \tag{1.10}$$

by means of the critical damping factor c_c. The solution to Eq. (1.8) describes the natural response of the vibratory system in the absence of the external forcing.

Depending on whether the critical damping factor is less than, equal to, or larger than 1, the vibrations are called, respectively, underdamped, critically damped, or overdamped.

1.2.2 Forced response — the resonance

When a force defined as

$$f(t) = F\sin(\omega t) \tag{1.11}$$

acts on the mass shown in Fig. 1.6, then Eq. (1.3) changes to

$$m\ddot{x} + c\dot{x} + kx = f(t) \tag{1.12}$$

The general solution of Eq. (1.12) is the sum of a complementary solution (which describes the system's vibration at the natural frequency) and a particular solution (which is vibration-generated at the driving frequency). The latter part of the solution is also called the steady-state solution and is generally analyzed in the frequency domain by studying its amplitude and phase angle.

Often Eq. (1.12) is written in the alternate form:

$$\ddot{x} + 2\xi\omega_r \dot{x} + \omega_r^2 x = \frac{F}{m}\sin(\omega t) \tag{1.13}$$

The solution to Eqs. (1.12) and (1.13), as shown by Timoshenko,[1] Thomson,[2] or Rao,[3] is the sum of the homogeneous solution—Eq. (1.8) —and a particular solution which is of the form:

$$x_p(t) = X\sin(\omega t - \varphi) \tag{1.14}$$

where the amplitude X is

$$X = \frac{X_{st}}{\sqrt{(1 - m\omega^2/k)^2 + (c\omega/k)^2}} = \frac{X_{st}}{\sqrt{(1 - \beta^2)^2 + (2\xi\beta)^2}} \tag{1.15}$$

with the frequency ratio β being defined as

$$\beta = \frac{\omega}{\omega_r} \tag{1.16}$$

and the phase angle between excitation and response φ as

$$\varphi = \arctan\frac{2\xi\beta}{1 - \beta^2} \tag{1.17}$$

The particular solution of Eq. (1.14) is of special importance as it describes the forced response of a vibratory system. In Eq. (1.15) the static displacement is X_{st} and is defined as F/k. Figures 1.7 and 1.8 are plots of the amounts X/X_{st} and φ as functions of β for various values of ξ.

As Fig. 1.7 indicates, when the driving frequency equals the resonant frequency ($\beta = 1$), the amplitude ratio reaches a maximum, which, for very small damping ratios, goes to infinity. Even in the presence of moderate damping, the amplitude at resonance is large, and this feature is utilized as a working principle in mechanical microresonators.

At resonance, when $\beta = 1$, the amplitude ratio of Eq. (1.15) becomes

$$\frac{X_r}{X_{st}} = \frac{1}{2\xi} \tag{1.18}$$

which gives an amplitude of

$$X = \frac{F_0}{2k\xi} \tag{1.19}$$

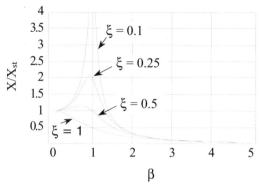

Figure 1.7 Plot of the amplitude ratio versus the frequency ratio.

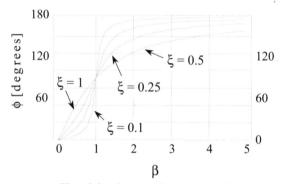

Figure 1.8 Plot of the phase angle versus the frequency ratio.

An important qualifier of mechanical microresonators is the quality factor Q, which, for a harmonic oscillator, is defined as

$$Q = \frac{2\pi\, U_s}{U_d} \tag{1.20}$$

where U_s is the energy stored and U_d is the energy dissipated during one cycle of oscillation. High quality factors indicate low losses through damping, and definitely microresonators having quality factors as large as possible are sought.

For a single-DOF damped system, such as the one pictured in Fig. 1.6, the energy stored per cycle is simply

$$U_s = \frac{k\,X^2}{2} \tag{1.21}$$

The energy dissipated per cycle is defined as

$$U_d = \int F_d \, dx \tag{1.22}$$

which, for viscous damping where the damping force is proportional to the oscillator's velocity:

$$F_d = c\dot{x} = \omega X \cos(\omega t - \varphi) \tag{1.23}$$

becomes

$$U_d = c \int_0^{2\pi/\omega} \dot{x}^2 \, dt = \pi c \omega X^2 \tag{1.24}$$

At resonance, the energy lost through damping is

$$U_{d,r} = 2\pi\xi k \, X^2 \tag{1.25}$$

By combining Eqs. (1.21) and (1.24), the quality factor of Eq. (1.20) becomes

$$Q = \frac{k}{c\omega} = \frac{1}{2\beta\xi} \tag{1.26}$$

At resonance, the quality factor is expressed as

$$Q_r = \frac{\sqrt{km}}{c} = \frac{1}{2\xi} \tag{1.27}$$

The resonance quality factor is also called *sharpness* at resonance (mostly in the mechanical vibration language, see Thomson[2] or Rao[3]), which is defined as the ratio $(\omega_2 - \omega_1)/\omega_r$, where the frequency difference in the numerator is

$$\omega_2 - \omega_1 = 2\xi\omega_r \tag{1.28}$$

These frequency values are also called sidebands or half-power points. It can be shown that this particular situation leads to an amplitude ratio of

$$\frac{X}{X_{st}} = \frac{1}{2\sqrt{2\xi}} \cong \frac{0.707}{2\xi} \tag{1.29}$$

and the case is pictured in the plot of Fig. 1.9.

The frequency difference $\omega_2 - \omega_1$ is called bandwidth being denoted by ω_b, and by combining Eqs. (1.25), and (1.29) and Fig. 1.9, the resonant Q factor can be expressed as

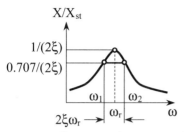

Figure 1.9 Sharpness of resonance with sidebands.

$$Q_r = \frac{1}{2\xi} = \frac{\omega_r}{\omega_2 - \omega_1} = \frac{\omega_r}{\omega_b} \qquad (1.30)$$

Large values of the Q factor microresonators require, as Eq. (1.30) indicates, high resonant frequencies and small bandwidths, which means small damping coefficients. It should also be mentioned that the bandwidth describes the ability of a resonant system to follow a sinusoidal driving signal which is close to the resonant frequency, and the bandwidth is proportional to the speed of response (see Ogata[4]).

Equation (1.27) also indicates that the quality factor can be defined for resonance as

$$Q_r = \frac{X_r}{X_{st}} \qquad (1.31)$$

Figure 1.10 plots the quality factor of Eq. (1.26) in terms of the frequency ratio for various damping ratios.

An alternative to qualifying the damping losses in mechanical resonators is the loss coefficient η (see Thomson,[2] for instance), which is the inverse of the quality factor:

$$\eta = \frac{1}{Q} = \frac{c\omega}{k} = 2\beta\xi \qquad (1.32)$$

The resonance loss coefficient is obviously

$$\eta_r = \frac{1}{Q_r} = 2\xi \qquad (1.33)$$

It is interesting to point out that the damping energy ratio can be expressed as

$$\frac{U_d}{U_{d,r}} = \beta \qquad (1.34)$$

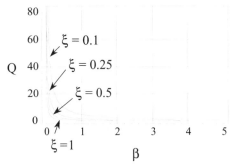

Figure 1.10 Quality factor as a function of the frequency ratio.

The energy loss mechanisms that are connected to the operation of mechanical microresonators are discussed next.

1.2.3 Loss mechanisms in mechanical microresonators

Energy loss phenomena in microdevices can generally be grouped into two large categories: One group includes losses that are produced by fluid-structure interaction, and the other group contains loss mechanisms that are generated through intrinsic (material) dissipation. Each category is briefly characterized in this section.

Fluid-structure interaction losses. One source of energy losses in NEMS/MEMS* is the interaction between a moving part and a fixed one, as fluid (air or liquid) is usually present between the two bodies in relative motion (except for the case where the oscillations take place in vacuum). Figure 1.11 schematically presents the main types of fluid-structure interaction damping.

In the sketch of Fig. 1.11a, the mobile plate moves against the fixed plate (the gap measured by the z coordinate is decreasing), and the result is the squeezing of the fluid film filling the variable gap, whence the name *squeeze-film damping*. The mobile plate in Fig. 1.11b moves parallel to the fixed plate by keeping the distance z_0 constant, and the effect on the interlaying film is one of shear.

In squeeze-film damping, the interaction among pressure, motion distances, time, plate geometry, and fluid film properties is governed by the Navier-Stokes partial differential equations. These equations simplify in the case of microdevices to the following equation:

---

* Nanoelectromechanical systems/microelectromechanical systems.

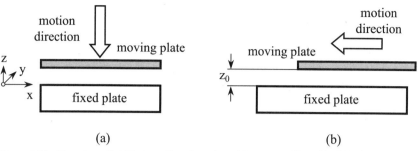

Figure 1.11   Structure-fluid interaction damping: (a) squeeze-film; (b) slide-film.

$$\frac{\partial^2 p}{\partial x^2} + \frac{\partial^2 p}{\partial y^2} = \frac{12\mu}{z_0^3}\frac{\partial z}{\partial t} \tag{1.35}$$

Equation (1.35), where p is the pressure in the fluid film and μ is the dynamic viscosity coefficient, is a Poisson-type equation which can be integrated (particularly when the contacting area of the two plates is elementary, such as rectangular or circular) and the pressure is the solution to it. Equation (1.35) is only valid, as mentioned by Starr,[5] when the Reynolds number Re satisfies the condition

$$Re = \frac{\rho\omega z^2}{\mu} \ll 1 \tag{1.36}$$

The Reynolds number is proportional to the ratio of inertia (turbulence) to viscosity, and therefore Eq. (1.36) requires that viscosity dominate. In addition to this Re number inequality, several other conditions need to be complied with in order for the Poisson equation, Eq. (1.35), to be valid, and these conditions and/assumptions are enumerated next. The fluid film should be in isothermal condition (which ensures that $\rho \approx p$ — and this simplifies the original Navier-Stokes equations); the pressure variations within the film are assumed to be small compared to the average pressure; the film thickness is uniform; the displacements by the movable plate are small compared to the film thickness; and the fluid is incompressible.

As mentioned previously, by integrating Poisson's equation for pressure and then determining the damping force (as a viscous one which is equal to the damping coefficient times the velocity), the damping coefficient can be obtained for a specified surface geometry. In the case of a circular conjugate surface, as mentioned by Starr[5] and by Andrews, Harris, and Turner,[6] the damping coefficient is

$$c = \frac{3\pi\mu R^4}{2z_0^3} \tag{1.37}$$

where R is the radius of the disk.

For a rectangular superposition area, the damping coefficient is also based on evaluations by Starr[5] and Andrews, Harris, and Turner,[6] namely,

$$c = \left(0.997 - 0.752\frac{w}{l} + 0.163\frac{w^2}{l^2}\right)\frac{w^3 l\mu}{z_0^3} \tag{1.38}$$

where w is the plate width and l is the plate length. For a very narrow strip, where $w/l \to 0$, Eq. (1.38) reduces to

$$c = 0.997\frac{w^3 l\mu}{z_0^3} \tag{1.39}$$

By knowing the damping coefficient c, it is possible to express the quality factor Q, which is connected to squeeze-film damping, according to Eq. (1.26):

$$Q = \frac{k}{c\omega} \tag{1.40}$$

where k is the stiffness of the mobile structure and ω is the relative motion frequency.

All previous development was based on the condition that the flow be a continuum. However, when the dimensions of the particles in the flow approach the relevant dimensions of the channel they travel in, the continuum property may no longer be valid. A quantifier that monitors this aspect is *Knudsen's number* Kn, which is defined as the ratio of the free mean molecular path to the relevant dimension of the channel, in this case:

$$\text{Kn} = \frac{\lambda}{z} \tag{1.41}$$

When Kn < 0.01, the continuum property of the flow is preserved; but when Kn < 10, the free mean path is comparable to (even larger than) the relevant channel dimension and the flow is free molecular. For the in-between range of 0.01 to 10 Kn, the flow is of a transition type where slip is possible. The Knudsen number can also be utilized as a correction factor in expressing the dynamic viscosity η variability as indicated by Veijola, Kuisma, and Lahdenpera,[7] who gave an effective value of

$$\mu_{eff} = \frac{\mu}{1 + 9.638 \text{Kn}^{1.159}} \qquad (1.42)$$

which was shown to be accurate for $0 < \text{Kn} < 880$.
A concern that is directly related to squeeze-film damping is the possibility that the fluid layer behaves as an elastic film when film compressibility becomes a noticeable factor. The predictor that indicates the possible spring behavior is the squeeze number, which, as shown by Starr,[5] is defined as

$$\sigma = \frac{12 \mu \omega l^2}{z_0^2 p_a} \qquad (1.43)$$

where l is the relevant in-plane dimension of the movable plate and p_a is the ambient fluid pressure. For squeeze numbers smaller than 0.2, as also indicated by Starr,[5] the film is practically incompressible and the spring behavior is negligible. For squeeze numbers larger than the 0.2 threshold value, the film behaves as a spring and therefore its energy dissipation properties are diminished. An even more relaxed condition is proposed by Blech[8] who mentions that for $\sigma < 3$ the gas escapes from the gap that is formed between the movable and fixed members, and there is no sensible gas compression; whereas for $\sigma > 3$, the gas is trapped in the gap and its compressibility generates the spring behavior. Starr[5] gave the stiffness of the film acting as a spring and also presented correction functions that have to be applied when the displacement of the mobile component is comparable to the film thickness.

In slide film damping (schematically shown in Fig. 1.11b), as indicated by Tang, Nguyen, and Howe[9]; Cho, Pisano, and Howe[10]; or Zhang and Tang,[11] there are a few flow regimes above and underneath the moving plate. Figure 1.12 shows the side view of a plate of mass m, which is attached by a spring of constant k. The mobile plate can be the finger of a comb drive microtransducer case in which the fixed plate underneath is the substrate. The distance between the two plates is constant and equal to z_0. When the plate oscillates with a frequency ω in the direction shown in the figure, a damping force (of viscous nature) resulting from the fluid-structure interaction will oppose the plate motion. There are basically two types of damping forces being generated by the following flows: a Couette-type flow, which is set between a mobile plate and a fixed one and where the velocity decays from v (the mobile plate velocity) at the mobile plate–fluid interface to zero at the fixed plate–fluid interface; and a Stokes-type flow, which is set both above and underneath the mobile plate. This flow is turbulent up to a

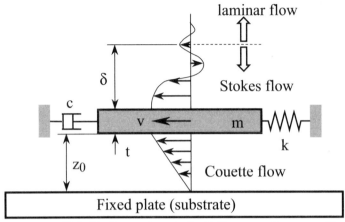

Figure 1.12 Slide-film flow regimes in the fluid surrounding a movable plate–fixed plate couple (side view).

distance δ (called the penetration distance) and becomes stationary past that distance. The penetration distance is calculated as

$$\delta = \sqrt{\frac{2\mu}{\rho\omega}} \qquad (1.44)$$

and represents the distance corresponding to the position where the velocity becomes

$$v_\delta = \frac{v}{e} \qquad (1.45)$$

where e is the natural logarithm base.

Figure 1.13 is the top view of the movable plate shown in relationship with a similar fixed plate, which is located at the same height z_0 from the substrate. The fixed and movable plates of Fig. 1.13 are the corresponding fingers in a comb microdevice, which is utilized for electrostatic transduction.

With reference to Fig. 1.12, the Couette flow damping can be quantified by the following quality factor, as shown by Cho, Pisano, and Howe:[10]

$$Q_{C,u} = \frac{\pi}{\omega}v^2\frac{\mu}{z_0} \qquad (1.46)$$

The subscript u indicates the zone underneath the movable plate. Similarly, and with reference to Fig. 1.13, the Couette flow quality factor

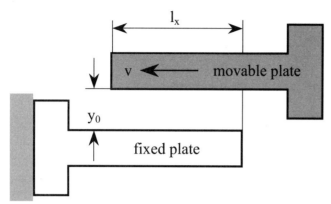

Figure 1.13 Slide-film flow in the fluid surrounding a movable plate–fixed plate couple (top view).

corresponding to the fluid damping generated through the relative motion on the side faces between two comb fingers is

$$Q_{C,s} = \frac{\pi}{\omega} v^2 \frac{\mu}{y_0} \tag{1.47}$$

where the subscript s points to the side region of the conjugate plates.

Stokes-type flow is responsible for damping produced both above and underneath the movable plate shown in Fig. 1.12. The turbulent Stokes flow which extends over the penetration height δ is characterized by the quality factor

$$Q_{S,a} = \frac{\pi}{\omega} v^2 \frac{\mu}{\delta} \tag{1.48}$$

According to Cho, Pisano, and Howe,[10] the Stokes flow damping underneath the moving plate is

$$Q_{S,u} = \frac{\pi}{\omega} v^2 \frac{\mu}{\delta} f(\beta) \tag{1.49}$$

where

$$f(\beta) = \frac{\sinh(2\beta z_0) + \sin(2\beta z_0)}{\cosh(2\beta z_0) - \cos(2\beta z_0)} \tag{1.50}$$

and

$$\beta = \frac{1}{\delta} \tag{1.51}$$

By collecting Eqs. (1.46), (1.47), (1.48), and (1.49), the overall quality factor that is produced by both Couette- and Stokes-type flows is calculated as

$$
\frac{1}{Q} = \frac{1}{Q_{C,u}} + \frac{1}{Q_{C,s}} + \frac{1}{Q_{S,a}} + \frac{1}{Q_{S,u}}
$$
$$
= \frac{\omega}{\pi v^2 \mu} \{ z_0 + y_0 + [1 + f(\beta)]\delta \}
$$

$$(1.52)$$

The quality factor is calculated in the particular addition form of Eq. (1.52) because it is inversely proportional to various energy loss sources, and the total damping energy is the sum of all the partial damping energies, such as the ones given in Eqs. (1.46) through (1.49).

Example: Compare the contributions of the four types of flows present in the slide-film phenomenon pictured in Figs. 1.12 and 1.13. Consider that the moving plate has the dimensions $l = 200$ μm, $w = 50$ μm, and $t = 2$ μm; and its mass density is $\rho = 2300$ kg/m^3. The spring stiffness is $k = 3.125$ N/m, and the fluid is air with a dynamic viscosity $\mu = 0.0345$ kg/ms.

The resonant frequency of the mass-spring system is 260,643 Hz, as calculated by Eq. (1.5). Also, the parameter β is 93,210. The following ratios can be formulated by Eqs. (1.46) through (1.49):

$$
\frac{1/Q_{C,u}}{1/Q} = \frac{Q}{Q_{C,u}} = \frac{z_0}{z_0 + y_0 + [1 + f(\beta)]\delta}
$$

$$
\frac{1/Q_{C,s}}{1/Q} = \frac{Q}{Q_{C,s}} = \frac{y_0}{z_0 + y_0 + [1 + f(\beta)]\delta}
$$

$$(1.53)$$

$$
\frac{1/Q_{S,a}}{1/Q} = \frac{Q}{Q_{S,a}} = \frac{\delta}{z_0 + y_0 + [1 + f(\beta)]\delta}
$$

$$
\frac{1/Q_{S,u}}{1/Q} = \frac{Q}{Q_{S,u}} = \frac{\delta f(\beta)}{z_0 + y_0 + [1 + f(\beta)]\delta}
$$

Equations (1.53) actually indicate that the overall quality factor is smaller than each of its components, since adding more damping factors increases the overall damping and consequently diminishes the overall quality factor. These ratios depend on only z_0 and y_0 and are plotted in Figs. 1.14 through 1.17.

The smallest contribution to the reduction of the overall quality factor is produced by the Stokes-type flow underneath the movable plate (Fig. 1.17), whereas the largest contribution in the overall quality factor reduction is generated by the Couette-type flow underneath the movable plate (Fig. 1.14) and the Stokes-type flow above the movable plate (Fig. 1.16).

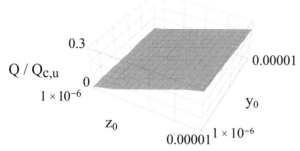

$Q / Q_{c,u}$

Figure 1.14 Quality factor contribution by the Couette flow underneath the movable plate.

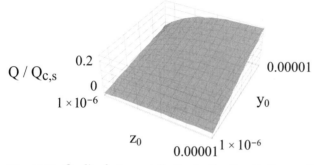

$Q / Q_{c,s}$

Figure 1.15 Quality factor contribution by the Couette flow on the side area.

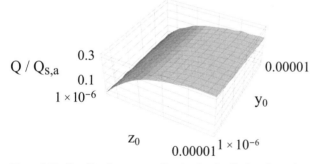

$Q / Q_{s,a}$

Figure 1.16 Quality factor contribution by the Stokes flow above the movable plate.

Internal (material) dissipation mechanisms. Besides losses in a mechanical microsystem that are produced through fluid-structure interaction and are of a viscous-damping nature, and other similar losses that are caused by friction between structures in relative motion (and that are not detailed here), another source of energy losses is represented by the internal dissipation mechanisms. These mechanisms, also referred to

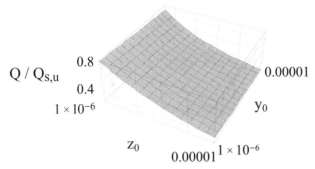

Figure 1.17 Quality factor contribution by the Stokes flow underneath the movable plate.

as mechanical noise mechanisms, are particularly important when the microdevice operates in vacuum, and therefore the energy dissipation through viscous damping is minimized. A global estimator of the mechanical noise in mechanical microoscillators is the *total noise equivalent acceleration* (TNEA) (see Yazdi, Ayazi, and Najafi[12]), defined as

$$\text{TNEA} = 2\frac{\sqrt{k_b T c}}{m} = 2\sqrt{\frac{k_b T \omega_r}{Qm}} \tag{1.54}$$

where k_b is the Boltzmann constant and T is the absolute temperature.

One source of internal energy losses in microresonators is the *thermoelastic energy dissipation* (TED) mechanism. More details on this temperature-related loss mechanism are given in the works of Roszhart[13] and Lifshitz and Roukes.[14] Roszhart[13] mentions that this mechanism is marked for beam resonators as thick as 10 μm, whereas Yasumara et al.[15] report experimental data showing that TED is significant down to beam thicknesses of 2.3 μm. For a bent beam, such as the one sketched in Fig. 1.18, the lower fibers are in tension and are cooler, whereas the upper fibers, which are in compression, are warmer than the undeformed beam. As a consequence of this temperature difference, a temperature gradient is set over the beam thickness which generates energy flow in the opposite direction, as shown in Fig. 1.18, and this mechanism generates irreversible energy losses in the mechanical microresonator.

The TED depends on material properties such as the coefficient of thermal expansion α, specific heat c_p, thermal conductivity κ, specific mass (density) ρ, elastic modulus (Young's modulus for bending) E, as well as on the temperature T and geometry. The quality factor which is related to TED can be quantified into the following form (after Roszhart[13]):

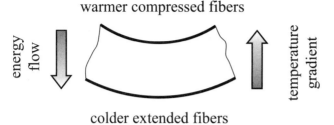

$$Q_{\text{TED}} = \frac{1.57\kappa}{E\alpha^2 T f_r t^2} \left(1 + \frac{0.4c_p^2 \rho^2 f_r^2 t^4}{\kappa^2}\right) \tag{1.55}$$

where f_r ($f_r = 2\pi\omega_r$) is the resonant frequency and t is the cross-sectional thickness. For a microcantilever (a fixed-free beam) Eq. (1.55) simplifies to

$$Q_{\text{TED}} = \frac{1.55\kappa l^2}{E\alpha^2 T t^2} \sqrt{\frac{\rho}{E}} \left(1 + \frac{0.42c_p^2 E \rho t^6}{\kappa^2 l^4}\right) \tag{1.56}$$

where l is the cantilever length.

Example: Analyze the TED quality factor for a silicon nitride microcantilever with $E = 126$ GPa, $\rho = 3440$ kg/m^3, $\alpha = 3 \times 10^{-6}$ K^{-1}, $c_p = 710$ J/(kg·K), and $\kappa = 3.2$ W/(m·K) for a temperature of 300 K.

Figure 1.19 highlights the fact that the quality factor owing to thermoelastic losses registers a minimum for thicknesses in the range of 5 to 20 µm. For thicknesses that are outside that range, the quality factor is not so severely affected by the TED mechanism. Figure 1.20 is another (two-dimensional) plot showing the temperature influence on the quality factor for a silicon nitride nanocantilever defined by a thickness $t = 100$ nm and a length $l = 60$ µm.

The vast majority of microresonators are fabricated as monolithic systems with virtually no friction losses due to the mating parts, losses which are present in macroscale rotation bearings, for instance. However, losses are still produced by energy dissipation through the anchors that attach the resonator to the substrate. Hosaka, Itao, and Kuroda,[16] as well as Yasumara et al.,[15] mention that the quality factor which expresses the clamping losses for cantilevers can be determined by analyzing the vibrational energy which is transmitted from the cantilever to the substrate (viewed as an infinite elastic plate) and gives an evaluation of

$$Q_{\text{cl}} \approx 2.17 \left(\frac{l}{t}\right)^3 \tag{1.57}$$

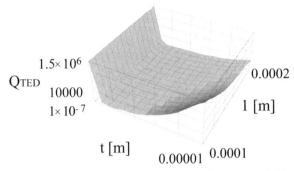

Q_{TED}

Figure 1.19 TED quality factor in terms of cantilever thickness and length.

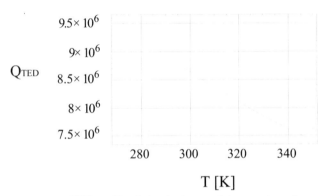

Q_{TED}

T [K]

Figure 1.20 TED quality factor in terms of temperature for a silicon nitride nanocantilever 100 nm thick and 60 μm thick.

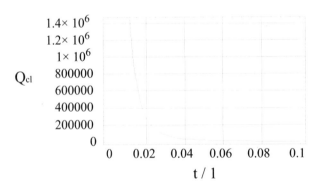

Q_{cl}

t / 1

Figure 1.21 Quality factor related to clamping losses as a function of the thickness-to-length ratio.

where l is the length and t is the thickness of a cantilever beam. Figure 1.21 is the two-dimensional plot illustrating this loss mechanism.

Intrinsic losses in the resonator material are important mechanisms accounting for energy dissipation. In a recent paper, Czaplewski et al.[17] analyzed the loss mechanisms in tetrahedral amorphous carbon (taC) and concluded that paramount in extrinsic internal dissipation is the defect motion, which is generated by structural reconfiguration through atomic motion between equilibrium or metastable states. The nomenclature and model of the anelastic or standard solid are utilized to quantify the material losses; see Freudenthal.[18] The model basis is the incompressible viscoelastic equation (also known as *Zener's model*, see Cleland and Roukes[19]), which describes such a behavior, namely, where

$$\sigma + \tau_1 \frac{d\sigma}{dt} = E(\varepsilon + \tau_2 \frac{d\varepsilon}{dt}) \qquad (1.58)$$

where σ = normal stress

ε = normal strain

τ_1 = stress-relaxation time

τ_2 = strain retardation time

This material model captures both the stress relaxation phenomenon (which means decrease of the stress in a component when constant strain is applied) and the strain retardation phenomenon, or creep (which implies the increase of strain under the application of a constant stress). Cleland and Roukes[19] suggested the following quality factor as being responsible for intrinsic losses:

$$\frac{1}{Q_m} = \frac{\omega\tau}{1 + \omega^2\tau^2} \frac{E_d - E}{E} \qquad (1.59)$$

where E_d is the dynamic (or unrelaxed) Young's modulus, which is experimentally determined by means of rapidly applying the test load, and σ is an aggregate relaxation time, which is calculated as

$$\tau = \sqrt{\tau_1\tau_2} \qquad (1.60)$$

Other factors accounting for energy dissipation in microresonators include losses through thin layers (surface losses), losses connected to the shape of the mechanical components, generated through phonon-phonon scattering, Nyquist-Johnson noise produced by electronic circuitry, transduction losses, adsorption-desorption processes from residual gases, environmental thermal drifts, or material defects such as those encountered at grain boundaries. By adding the losses that have been discussed in this section, the overall quality factor is calculated as

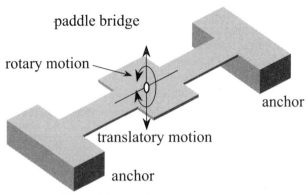

paddle bridge

rotary motion

anchor

translatory motion

anchor

Figure 1.22 Microbridge as a 2-DOF system.

$$\frac{1}{Q_{overall}} = \sum_i \frac{1}{Q_i} \tag{1.61}$$

where the Q_i are all identifiable and quantifiable individual quality factors.

1.3   Multiple-Degree-of-Freedom Systems

Mechanical microresonators that are capable of more than one motion are modeled as multiple-DOF members. A paddle microbridge, such as the one shown in Fig. 1.22 and which can be used at detecting extraneous substances through shifts in the resonant frequencies (more details on mass addition detection are offered in Chap. 6), may vibrate in bending and in torsion, and therefore it needs to be modeled as a 2-DOF system.

This section first discusses the approximation methods of Rayleigh and Dunkerley that permit evaluation of the upper and lower bounds on the resonant frequencies of a multiple-DOF system. Presented next are the notions of eigenvalues, eigenvectors, and eigenmodes (or mode shapes) as well as the static and/dynamic coupling. Lagrange's equations are studied subsequently as a tool of formulating the equations of motion of a vibrating system. Mechanical-electrical analogies are presented in the end with the main notions of the Laplace transform, the transfer function, and a mechanical resonator filter application.

1.3.1   Approximate methods for resonant frequencies calculation

In many instances only the extreme resonant frequencies are relevant in the design of a microresonator which is modeled as a multiple-DOF system. Two methods are presented next that enable calculation of the maximum and minimum resonant frequencies without requiring calculation of all the resonant frequencies defining the multiple-DOF system.

Rayleigh's method and the upper bound on resonant frequencies by the stiffness method. For a multiple-DOF system, the dynamic equation defining the free undamped vibrations is

$$[M]\{\ddot{x}\} + [K]\{x\} = 0 \tag{1.62}$$

where [M] is the mass matrix and [K] is the stiffness matrix. In a harmonic motion, the acceleration vector can be expressed in terms of the displacement vector and the frequency as

$$\{\ddot{x}\} = -\omega^2\{x\} \tag{1.63}$$

By combining Eqs. (1.62) and (1.63), the following equation can be written:

$$([M]^{-1}[K] - \omega^2[I])\{x\} = 0 \tag{1.64}$$

The characteristic equation corresponding to Eq. (1.64) is

$$\det([M]^{-1}[K] - \omega^2[I]) = 0 \tag{1.65}$$

and by solving it the resonant frequencies ω are obtained. The method of finding the resonant frequencies and the corresponding modes based on the stiffness matrix is called the stiffness method. In general, the characteristic equation can be written as

$$-\omega^{2n} + (a_{11} + a_{22} + \cdots + a_{nn})\omega^{2(n-1)} + \cdots = 0 \tag{1.66}$$

where $a_{11}, a_{22}, \ldots, a_{nn}$ are the diagonal terms of the dynamic matrix, which is defined as

$$[A] = [M]^{-1}[K] \tag{1.67}$$

Related to Eq. (1.66), it is well known from algebra that

$$\omega_1^2 + \omega_2^2 + \cdots + \omega_n^2 = a_{11} + a_{22} + \cdots + a_{nn} \tag{1.68}$$

where $\omega_1, \omega_2, \ldots, \omega_n$ are the resonant frequencies. It follows from Eq. (1.68) that

$$\omega_1^2 < a_{11} + a_{22} + \cdots + a_{nn} \tag{1.69}$$

In other words, the upper bound on the first (natural) frequency of an n-DOF system is the sum of the main diagonal terms; that sum is also known as the *trace* of the dynamic matrix. Equation (1.69) is the mathematical formulation of Rayleigh's procedure.

Dunkerley's method and the lower bound on resonant frequencies by the compliance method. The lower bound on the resonant frequencies is determined by applying Dunkerley's method, which is presented next. The inertia force acting on the undamped dynamic system formed of n masses is

$$\{F\} = -[M]\{\ddot{x}\} \tag{1.70}$$

At the same time, the displacement vector can be expressed as

$$\{x\} = [C]\{F\} \tag{1.71}$$

where $[C]$ is the compliance matrix (the inverse of the stiffness matrix $[K]$). By combining Eqs. (1.62), (1.63), (1.67), and (1.71), the following equation is obtained:

$$\left([C][M] - \frac{[I]}{\omega^2}\right)\{x\} = 0 \tag{1.72}$$

which is also known as the *modal equation* formulated by means of the flexibility or compliance method. The characteristic equation corresponding to Eq. (1.72) is

$$\det\left([C][M] - \frac{[I]}{\omega^2}\right) = 0 \tag{1.73}$$

The dynamic matrix according to the compliance approach is defined as

$$[A] = [C][M] \tag{1.74}$$

and it can be shown that

$$\frac{1}{\omega_1^2} + \frac{1}{\omega_2^2} + \cdots + \frac{1}{\omega_n^2} = C_{11}M_1 + C_{22}M_2 + C_{nn}M_n \tag{1.75}$$

where the C_{ii} are the diagonal terms of the symmetric compliance matrix and the M_i are the terms of the diagonal mass matrix. Equation (1.75) enables us to state that

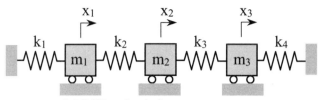

Figure 1.23   A 3-DOF mechanical resonator.

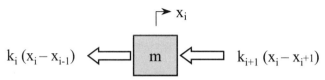

$$k_i\,(x_i - x_{i\text{-}1}) \Longleftarrow \boxed{m} \Longleftarrow k_{i+1}\,(x_i - x_{i+1})$$

Figure 1.24   Free-body diagram of one body in the 3-DOF system of Fig. 1.23.

$$\omega_1^2 > \frac{1}{C_{11}M_1 + C_{22}M_2 + \cdots + C_{nn}M_n} \tag{1.76}$$

Equation (1.76) gives the lower bound of the natural frequency of a n-DOF system, by means of the Dunkerley's method.

Example: Determine the upper and lower bounds on the resonant frequencies for the 3-DOF system of Fig. 1.23 which models a mechanical resonator filter.

The equation of motion for body i (i = 1, 2, 3) can be found by applying Newton's second law, for instance,

$$m_i \ddot{x}_i = -k_{i+1}(x_i - x_{i+1}) - k_i(x_i - x_{i-1}) \tag{1.77}$$

and is based on the free-body diagram (FBD) of Fig. 1.24.

Equation (1.77) can be written in the matrix form of Eq. (1.62) where the mass matrix is

$$[M] = \begin{bmatrix} m_1 & 0 & 0 \\ 0 & m_2 & 0 \\ 0 & 0 & m_3 \end{bmatrix} \tag{1.78}$$

and the stiffness matrix is

$$[K] = \begin{bmatrix} k_1 + k_2 & -k_2 & 0 \\ -k_2 & k_2 + k_3 & -k_3 \\ 0 & -k_3 & k_3 + k_4 \end{bmatrix} \tag{1.79}$$

The stiffness-based dynamic matrix defined in Eq. (1.67) is

$$
[A] = \begin{bmatrix} \dfrac{k_1+k_2}{m_1} & -\dfrac{k_2}{m_1} & 0 \\[3mm] -\dfrac{k_2}{m_2} & \dfrac{k_2+k_3}{m_2} & -\dfrac{k_3}{m_2} \\[3mm] 0 & -\dfrac{k_3}{m_3} & \dfrac{k_3+k_4}{m_3} \end{bmatrix}
\tag{1.80}
$$

and therefore the upper bound on the natural frequencies is found according to Rayleigh's procedure [Eq. (1.69)] as

$$
\omega_u = \sqrt{\dfrac{k_1+k_2}{m_1} + \dfrac{k_2+k_3}{m_2} + \dfrac{k_3+k_4}{m_3}}
\tag{1.81}
$$

When all the masses and springs of the system are identical, Eq. (1.81) simplifies to

$$
\omega_u^* = \sqrt{6}\sqrt{\dfrac{k}{m}}
\tag{1.82}
$$

The compliance matrix, which is needed in Dunkerley's method for the lower natural frequency calculation, is the inverse of the stiffness matrix of Eq. (1.79). By using the diagonal terms of it, together with the masses m_1, m_2, and m_3, the lower resonant frequency, according to Eq. (1.76), is found to be

$$
\omega_l = \sqrt{\dfrac{k_1 k_2 k_3 + k_2 k_3 k_4 + k_1 k_4(k_2+k_3)}{k_1 k_4 m_2 + k_2(k_3+k_4)(m_1+m_2) + k_2(k_1+k_3)m_3 + k_3[k_4 m_1 + k_1(m_2+m_3)]}}
\tag{1.83}
$$

When the masses and springs are identical, Eq. (1.83) reduces to

$$
\omega_l^* = \sqrt{0.4}\sqrt{\dfrac{k}{m}}
\tag{1.84}
$$

1.3.2 Eigenvalues, eigenvectors, and mode shapes

It has been shown that the matrix equation governing the free undamped response of a lumped, multiple-DOF vibratory system has the following form, corresponding to the ith normal mode:

$$
([A] - \lambda_i[I])\{X_i\} = \{0\}
\tag{1.85}
$$

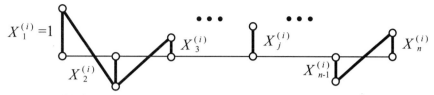

Figure 1.25   Mode shape i corresponding to an eigenvector of dimension n.

where $[A]$ is the dynamic matrix defined as in Eq. (1.67). The eigenvalue λ_i is connected to the resonant frequency as

$$\lambda_i = \omega_i^2 \tag{1.86}$$

and the vector of Eq. (1.85) is the eigenvector corresponding to the eigenvalue λ_i. The matrix form Eq. (1.85) is formed of n homogeneous algebraic equations under the assumption that the system's dimension is n. The unknowns in these equations are the n components of the eigenvector $\{X_i\}$. Because the equations are homogeneous, only $n - 1$ unknown eigenvector components can be determined, all in terms of one eigenvector component, which can be chosen arbitrarily. Usually, the arbitrary component is equal to 1 and the other components are less than 1, and this form can be achieved by a normalization procedure. The graphical representation of the components of one eigenvector is illustrated in Fig. 1.25 and is known as the *eigenmode* or *mode shape*. As a result, an n-DOF mechanical system has n eigenvalues, and for each eigenvalue there is an eigenvector and the corresponding mode shape.

The following relationship also holds true:

$$\det\left([A] - \lambda[I]\right)[I] = \left([A] - \lambda[I]\right)\mathrm{adj}([A] - \lambda[I]) \tag{1.87}$$

where for a square nonsingular matrix $[B]$, its adjoint is found by means of

$$[B]^{-1} = \frac{\mathrm{adj}[B]}{\det[B]} \tag{1.88}$$

For the ith mode, the determinant of Eq. (1.87) is zero; and therefore, by comparing Eqs. (1.85) and (1.87), it follows that one column of the adjoint matrix $([A] - \lambda[I])$ is the eigenvector corresponding to the ith mode. In other words, finding a specific eigenvector can be done by formulating the adjoint matrix for that mode.

It should be pointed out that normal modes are orthogonal with respect to the stiffness and mass matrices, namely,

$$\{X_i\}^T[M]\{X_j\} = 0 \qquad \{X_i\}^T[K]\{X_j\} = 0 \qquad (1.89)$$

for $i \neq j$. When $i = j$, the generalized stiffness and mass matrices can be defined as

$$M_i = \{X_i\}^T[M]\{X_i\} \qquad K_i = \{X_i\}^T[K]\{X_i\} \qquad (1.90)$$

The generalized mass and stiffness matrices of Eqs. (1.90) can be utilized to quantify the participation of various modes in defining the free vibrations of a multiple-DOF system as the superposition of normal modes in the form:

$$\{X\} = \sum_i c_i\{X_i\} \qquad (1.91)$$

where c_i is a participation factor corresponding to the ith mode. Premultiplication in Eq. (1.91) by $\{X_i\}^T[M]$ gives the mass participation factor as

$$c_i^M = \frac{\{X_i\}^T[M]\{X\}}{M_i} \qquad (1.92)$$

Similarly, the stiffness participation factor is

$$c_i^K = \frac{\{X_i\}^T[K]\{X\}}{K_i} \qquad (1.93)$$

It can be shown — see Thomson,[2] for instance — that by using the modal matrix, defined as

$$[P] = [\{X_1\}\{X_2\}\ldots\{X_i\}\ldots\{X_n\}] \qquad (1.94)$$

where $\{X_i\}$ is the ith eigenvector, the equation of motion can be decoupled, as it can be written in the form:

$$[M_m]\{\ddot{x}\} + [K_m]\{x\} = \{0\} \qquad (1.95)$$

where the transformed (modal) mass and stiffness matrices

$$[M_m] = [P]^T[M][P] \qquad [K_m] = [P]^T[K][P] \qquad (1.96)$$

are both of diagonal form.

For an n-DOF dynamic system, the mass matrix $[M]$ and the stiffness matrix $[K]$ are symmetric and generally fully populated. Such a system is also known as fully coupled. Systems where the mass matrix is in

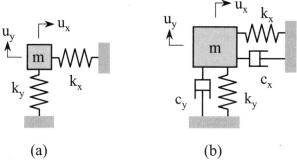

Figure 1.26 Model of a 2-DOF gyroscope: (*a*) without damping; (*b*) with damping.

diagonal form are mass- or dynamically decoupled, whereas systems where the stiffness matrix is in diagonal form are statically decoupled. A system is fully decoupled when both the mass and the stiffness matrices are in diagonal form. In many cases, the design effort is directed at realizing one or both of the decoupling forms. Also note that coupling and decoupling depend on the manner of selecting the coordinates.

Example: A gyroscope (this subject is treated in greater detail in Chap. 5), which is considered as a 2-DOF system under the assumption of small deformations, is studied now based on Fig. 1.26*a*.
   The equations of the undamped motion about the two directions are

$$m\ddot{x} + k_x x = 0 \quad m\ddot{y} + k_y y = 0 \tag{1.97}$$

These equations can be collected into the matrix form:

$$\begin{bmatrix} m & 0 \\ 0 & m \end{bmatrix} \begin{Bmatrix} \ddot{x} \\ \ddot{y} \end{Bmatrix} + \begin{bmatrix} k_x & 0 \\ 0 & k_y \end{bmatrix} = \begin{Bmatrix} 0 \\ 0 \end{Bmatrix} \tag{1.98}$$

which shows that the system is decoupled both statically and dynamically. Similarly, the damped motions about x and y are based on the sketch of Fig. 1.26*b* and can be written as

$$\begin{bmatrix} m & 0 \\ 0 & m \end{bmatrix} \begin{Bmatrix} \ddot{x} \\ \ddot{y} \end{Bmatrix} + \begin{bmatrix} c_x & 0 \\ 0 & c_y \end{bmatrix} \begin{Bmatrix} \dot{x} \\ \dot{y} \end{Bmatrix} + \begin{bmatrix} k_x & 0 \\ 0 & k_y \end{bmatrix} \begin{Bmatrix} x \\ y \end{Bmatrix} = \begin{Bmatrix} 0 \\ 0 \end{Bmatrix} \tag{1.99}$$

This system, too, is decoupled, both statically (because of the diagonal shape of the stiffness matrix) and dynamically (because of the diagonal mass and damping matrices).

Example: Study the coupling and/decoupling of the 3-DOF system shown in Fig. 1.23, and find the transformed matrices which will decouple the system in case it is coupled. Consider that $m_1 = m_2 = m_3 = m$ and $k_1 = k_2 = k_3 = k_4 = k$.

As Eq. (1.79) shows, the system is statically coupled as the stiffness matrix is not in diagonal form. The eigenvectors corresponding to the mechanical system of Fig. 1.23 are assembled into the modal matrix of Eq. (1.93) as

$$[P] = \begin{bmatrix} -1 & 1 & 1 \\ 0 & \sqrt{2} & -\sqrt{2} \\ 1 & 1 & 1 \end{bmatrix} \tag{1.100}$$

As a result of the definition in Eq. (1.96), the transformed stiffness matrix is

$$[K_m] = 4k \begin{bmatrix} 1 & 0 & 0 \\ 0 & 2 - \sqrt{2} & 0 \\ 0 & 0 & 2 + \sqrt{2} \end{bmatrix} \tag{1.101}$$

which is in diagonal form. The mass matrix [Eq. (1.78)] is already in diagonal form, and there is no need to calculate a transformed mass matrix. The decoupled dynamic equation of motion will therefore be

$$[M]\{\ddot{x}\} + [K_m]\{x\} = \{0\} \tag{1.102}$$

1.3.3 Lagrange's equations

Lagrange's method and equations can be utilized to derive the equations of motion of a mechanical system with (generally) multiple degrees of freedom by utilizing the energy forms that define the system's motion. Presented next are Lagrange's equations for conservative systems as well as for nonconservative ones.

In conservative systems there are no energy gains or losses, as indicated by Timoshenko,[1] Thomson,[2] or Rao,[3] and Fig. 1.27 shows a serial chain composed of only masses and springs. The system possesses n degrees of freedom. The number n represents the minimum number of parameters that can describe the state of the system at any given time, and these parameters (depending on time) are called generalized coordinates.

Lagrange's equation for the ith mass (and degree of freedom) is

$$\frac{d}{dt}\left(\frac{\partial T}{\partial \dot{x}_i}\right) - \frac{\partial T}{\partial x_i} + \frac{\partial U}{\partial x_i} = 0 \tag{1.103}$$

where T is the kinetic energy of the whole system and is calculated as

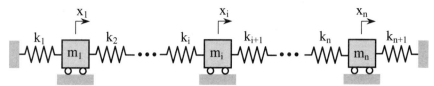

Figure 1.27 Conservative mass-spring system.

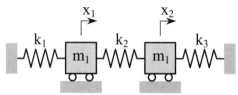

Figure 1.28 Two-microresonator filter as a 2-DOF system.

$$T = \sum_{i=1}^{n} \left(\frac{1}{2} m_i \dot{x}_i^2 \right) \tag{1.104}$$

and U is the elastic potential energy of the whole system and is calculated as

$$U = \sum_{i=1}^{n+1} \left[\frac{1}{2} k_i (x_{i+1} - x_i)^2 \right] \tag{1.105}$$

The kinetic energy can also be written in vector-matrix form as

$$T = \frac{1}{2} \{\dot{x}\}^T [M] \{\dot{x}\} \tag{1.106}$$

whereas the potential is similarly expressed as

$$U = \frac{1}{2} \{x\}^T [K] \{x\} \tag{1.107}$$

In Eqs. (1.106) and (1.107), $\{x\}$ is the vector of generalized coordinates, $[M]$ is the mass matrix, and $[K]$ is the stiffness matrix.

Example: Determine the resonant frequencies and the associated mode shapes (eigenvectors and eigenmodes) for the two-microresonator filter of Fig. 1.28. Consider the particular case where $m_1 = m$, $m_2 = 2m$, and $k_1 = k_2 = k_3 = k$.

The kinetic energy of the 2-DOF system of Fig. 1.28 is

$$T = \frac{1}{2} m_1 \dot{x}_1^2 + \frac{1}{2} m_2 \dot{x}_2^2 \tag{1.108}$$

and the potential elastic energy is

$$U = \frac{1}{2}k_1 x_1^2 + \frac{1}{2}k_2(x_2 - x_1)^2 + \frac{1}{2}k_3 x_2^2 \tag{1.109}$$

By applying the Lagrange equations approach, the following differential equations describing the free response of the system are obtained:

$$m_1\ddot{x}1 + (k_1 + k_2)x_1 - k_2 x_2 = 0 \quad m_2\ddot{x}2 + (k_2 + k_3)x_2 - k_2 x_1 = 0 \tag{1.110}$$

Equations (1.110) can also be written in vector-matrix form as

$$\begin{bmatrix} m_1 & 0 \\ 0 & m_2 \end{bmatrix} \begin{Bmatrix} \ddot{x}1 \\ \ddot{x}2 \end{Bmatrix} + \begin{bmatrix} k_1 + k_2 & -k_2 \\ -k_2 & k_2 + k_3 \end{bmatrix} \begin{Bmatrix} x_1 \\ x_2 \end{Bmatrix} = \begin{Bmatrix} 0 \\ 0 \end{Bmatrix} \tag{1.111}$$

The first matrix 0 in the left-hand side of Eq. (1.111) is the mass matrix, whereas the second matrix is the stiffness matrix. The dynamic matrix is

$$[A] = [M]^{-1}[K] = \begin{bmatrix} \dfrac{k_1 + k_2}{m_1} & -\dfrac{k_2}{m_1} \\ -\dfrac{k_2}{m_2} & \dfrac{k_2 + k_3}{m_2} \end{bmatrix} \tag{1.112}$$

The eigenvalues of this matrix are solutions to the characteristic equation, and by taking the particular values of this problem, the corresponding resonant frequencies are

$$\omega_1 = \sqrt{\frac{3 - \sqrt{3}}{2}\frac{k}{m}} \quad \omega_2 = \sqrt{\frac{3 + \sqrt{3}}{2}\frac{k}{m}} \tag{1.113}$$

The first mode is expressed by the following equation:

$$\begin{bmatrix} \dfrac{k_1 + k_2}{m_1} - \omega_1^2 & -\dfrac{k_2}{m_1} \\ -\dfrac{k_2}{m_2} & \dfrac{k_2 + k_3}{m_2} - \omega_1^2 \end{bmatrix} \begin{Bmatrix} X_1^{(1)} \\ X_2^{(1)} \end{Bmatrix} = \begin{Bmatrix} 0 \\ 0 \end{Bmatrix} \tag{1.114}$$

The first eigenvector is

$$\{X_1\} = \begin{Bmatrix} -1 + \sqrt{3} \\ 1 \end{Bmatrix} \tag{1.115}$$

whereas the second eigenvector, which corresponds to the second resonant frequency of Eq. (1.113), is

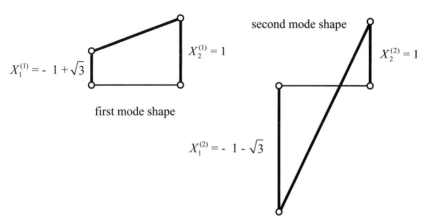

$X_2^{(1)} = 1$

$X_1^{(1)} = -1 + \sqrt{3}$

second mode shape

$X_2^{(2)} = 1$

first mode shape

$X_1^{(2)} = -1 - \sqrt{3}$

Figure 1.29 Mode shapes for the example of Fig. 1.28.

$$\{X_2\} = \left\{ \begin{array}{c} -1 - \sqrt{3} \\ 0 \end{array} \right\} \tag{1.116}$$

The two eigenmodes (or mode shapes) of this problem are sketched in Fig. 1.29.

For nonconservative systems, such as those where energy is added to or drained away from the system, the Lagrange equations are

$$\frac{d}{dt}\left(\frac{\partial T}{\partial \dot{x}_i}\right) - \frac{\partial T}{\partial x_i} + \frac{\partial U}{\partial x_i} = F_i \tag{1.117}$$

where F_i is the nonconservative generalized force corresponding to the ith degree of freedom.

1.4   Mechanical-Electrical Analogies for Microsystems

Microelectromechanical systems, as their name suggests, combine electrical and mechanical components into one system. Expressing the interaction between electrical and mechanical phenomena is simplified when the mechanical-electrical analogies are employed. In essence, as mentioned by Ogata[4] or Mazet,[20] two systems are analogous when they can be described by similar mathematical models. For instance, when the differential equation defining the behavior of a mechanical system has the same form as the equation governing the evolution of an electrical system, the two systems are analogous.

Two basic forms of mechanical-electrical analogies are discussed next, namely, the force-voltage (or mass-inductance) analogy and the

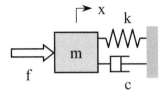

Figure 1.30   Single-DOF mechanical system.

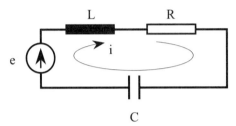

Figure 1.31   Series LRC electric circuit.

force-current (or mass-capacitance) analogy. Both variants utilize the same mechanical system, whereas the electrical counterpart is different.

Figure 1.30 illustrates a single-DOF system consisting of a mass connected to a spring and to a dashpot (spring-damper combination) and acted upon by an external force.

The differential equation describing the motion of the mass in Fig. 1.30 is

$$m \frac{d^2 x}{dt^2} + c \frac{dx}{dt} + kx = F \tag{1.118}$$

The series electric circuit of Fig. 1.31 is now considered, which is composed of a voltage source e, an inductor having inductance L, a resistor of resistance R, and a capacitor of capacitance C.

By applying Kirchhoff's second (circuit) law, the source voltage can be expressed as

$$e = L \frac{di}{dt} + Ri + \frac{1}{C} \int i \, dt \tag{1.119}$$

The current i is related to the electrical charge q as

$$i = \frac{dq}{dt} \tag{1.120}$$

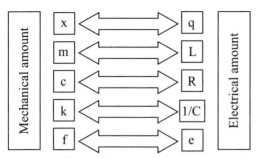

Figure 1.32   Force-voltage (mass-inductance) analogy.

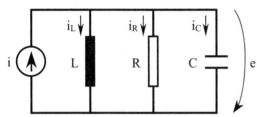

Figure 1.33   Parallel *LRC* electric circuit.

and therefore Eq. (1.119) can be written as

$$L \, \frac{d^2 q}{dt^2} + R \, \frac{dq}{dt} + \frac{1}{C} q = e \tag{1.121}$$

It can be seen that Eq. (1.118), describing the forced motion of the mechanical system sketched in Fig. 1.30, and Eq. (1.121), defining the time behavior of the electric circuit shown in Fig. 1.31, are both second-order differential equations with constant coefficients, and therefore the two systems are analogous. Figure 1.32 illustrates the direct relationships between the amounts defining the two systems.

The other analogy that can be drawn between a mechanical and an electrical system is also based on the single-degree-of-freedom system of Fig. 1.30 and the electrical system sketched in Fig. 1.33.

Kirchhoff's first (or node) law shows that the source current *i* is the sum of currents passing through the inductor, resistor, and capacitor:

$$i = i_L + i_R + i_C \tag{1.122}$$

By taking into account the current definitions

$$i_L = \frac{1}{L} \int C \, dt \quad i_R = \frac{e}{R} \quad i_C = C \, \frac{de}{dt} \tag{1.123}$$

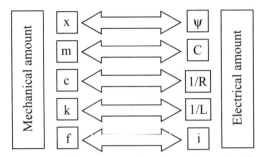

Figure 1.34   Force-current (mass-capacitance) analogy.

as well as the equation connecting the voltage e to the magnetic flux linkage ψ:

$$e = \frac{d\psi}{dt} \qquad (1.124)$$

Equation (1.122) can be reformulated as

$$C\frac{d^2\psi}{dt^2} + \frac{1}{R}\frac{d\psi}{dt} + \frac{1}{L}\psi = i \qquad (1.125)$$

The similarity between Eq. (1.125) describing the parallel LRC circuit and Eq. (1.118) which defines the behavior of the spring-mass-damper system of Fig. 1.30 can be noted, and therefore it can be concluded the two systems are analogous. The direct relationships between the corresponding amounts in the two systems are shown in Fig. 1.34.

1.5   Laplace Transforms, Transfer Functions, and Complex Impedances

A convenient tool for solving system dynamics and control problems which are encountered in modeling and designing NEMS/MEMS (particularly microresonators) is the Laplace transform, which is an operational method. The Laplace transform, as shown by Ogata,[4] for instance, is defined as

$$\mathscr{L}[f(t)] = F(s) = \int_0^\infty f(t)e^{-st}\,dt \qquad (1.126)$$

In essence, the Laplace transform takes a given function depending on time (for example) $f(t)$ and by means of the integral of Eq. (1.126) transforms that function into another function F depending on another variable s. The new function $F(s)$ is called the Laplace transform of the

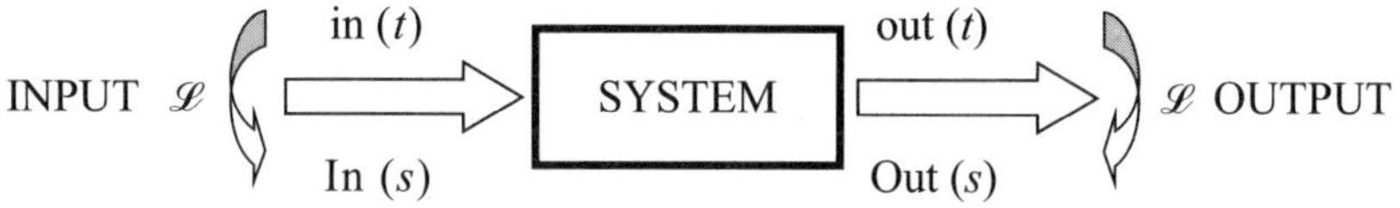

Figure 1.35 Simple system with input (driving) and output (response) functions.

original function $f(t)$. It is beyond the scope of this book to go into detail with the Laplace transform and its related aspects, but it should be mentioned briefly that this transform enables approaching and solving integral differential equations (such as the one introduced previously in defining the behavior of mechanical and electrical systems) that are transformed into algebraic equations in the Laplace s domain. The most usual Laplace transforms that are connected to differentiation and integration are

$$\mathscr{L}\left[\frac{d^2 f(t)}{dt^2}\right] = s^2 F(s) \quad \mathscr{L}\left[\frac{df(t)}{dt}\right] = sF(s) \quad \mathscr{L}\left[\int f(t)\,dt\right] = \frac{F(s)}{s} \quad (1.127)$$

where the first two of Eqs. (1.127) are valid when $f(0) = 0$ and $df(t)/dx = 0$ (for $t = 0$).

The Laplace transform is directly utilized in defining the transfer function of a system. Consider a system that is defined by specific properties and is acted upon by an input function (also called a driving function). The system input interaction results in an output function (or response function) as sketched in Fig. 1.35.

The system properties can be conveniently characterized by means of the transfer function TF, which is defined as the ratio of the Laplace transform of the output function to the Laplace transform of the input function:

$$\text{TF} = \frac{\mathscr{L}[\text{out}(t)]}{\mathscr{L}[\text{in}(t)]} = \frac{\text{Out}(s)}{\text{In}(s)} \quad (1.128)$$

Example: Determine the transfer function of the single-DOF system of Fig. 1.30 by considering Fig. 1.36, which indicates that the input function is the force $f(t)$, and the output is the displacement $x(t)$.

It has been shown that the time-domain behavior of the simple mechanical system is governed by the differential equation

$$m\frac{d^2 x(t)}{dt^2} + c\frac{dx(t)}{dt} + k\,x(t) = f(t) \quad (1.129)$$

By applying the Laplace transform to Eq. (1.129) and by taking into account the rules outlined in Eqs. (1.127), the following equation is obtained that describes the system behavior into the Laplace s domain:

Figure 1.36   Transfer function diagram of a mass-spring-damper system.

$$(ms^2 + cs + k)X(s) = F(s) \qquad (1.130)$$

Equation (1.130) enables us to express the transfer function of the mechanical system as

$$\mathrm{TF} = \frac{X(s)}{F(s)} = \frac{1}{ms^2 + cs + k} \qquad (1.131)$$

Recalling the mechanical-electrical analogies discussed previously in this section, we emphasize that a more mathematically sound definition of the analogy states that two different systems are analogous if they have the same transfer function (Ogata[4]), and this is an extremely important feature that allows unitary mathematical treatment of physically different systems. For instance, an analog of the mechanical system of Fig. 1.30 is the series electrical system sketched in Fig. 1.31. By following an approach similar to the one taken for the mechanical systems, it can be shown that the transfer function of the electrical system is

$$\mathrm{TF} = \frac{Q(s)}{E(s)} = \frac{1}{Ls^2 + Rs + 1/C} \qquad (1.132)$$

In other words, and strictly speaking, the mechanical and electrical systems are analogous when the following conditions are satisfied:

$$m \sim L \qquad c \sim R \qquad k \sim \frac{1}{C} \qquad (1.133)$$

There are microresonator systems, as will be shown shortly, that are formed as serial connections between simple subsystems (units), and this system is schematically illustrated in Fig. 1.37.

It is assumed that the transfer functions of all component subsystems TF_1, TF_2, ..., TF_n are known, and the aim is to find a global (system) transfer function to be equivalent to the component subsystems in a manner that will preserve the input Laplace-transformed $I(s)$ and output $O(s)$ functions. Figure 1.37 also indicates that the output from an intermediate subsystem is the input to the immediately following one, and therefore

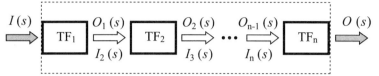

Figure 1.37 System formed of serially connected unit subsystems.

$$O_1(s) = I_2(s)$$
$$O_2(s) = I_3(s)$$
$$\vdots$$
$$O_{n-1}(s) = I_n(s)$$

(1.134)

which means that

$$TF_1(s) = \frac{O_1(s)}{I(s)} = \frac{I_2(s)}{I(s)}$$

$$TF_2(s) = \frac{O_2(s)}{I_2(s)} = \frac{I_3(s)}{I_2(s)}$$

$$\vdots$$

$$TF_n(s) = \frac{O(s)}{I_n(s)}$$

(1.135)

It can now easily be checked that

$$\frac{O(s)}{I(s)} = TF_1(s)\,TF_2(s)\ldots TF_n(s) = TF(s)$$

(1.136)

In other words, Eq. (1.136) indicates that the transfer function of the entire system, which is composed of the n serially connected subsystems, is equal to the product of all component transfer functions.

Another useful concept (amount) in describing microelectromechanical systems is the complex impedance; see Ogata,[4] for instance. Figure 1.38 defines the complex impedance $Z(s)$ as being the particular transfer function which connects the Laplace transform of the output voltage to the Laplace transform of the input current into an electric piece of circuit. This definition and the use of impedances enable us to unitarily treat inductors, resistors, and capacitors.

The complex impedance is therefore expressed as

$$Z(s) = \frac{E(s)}{I(s)}$$

(1.137)

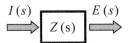

Figure 1.38 Complex impedance definition.

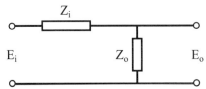

Figure 1.39 Typical input-output complex impedance-based system.

The complex impedance of a series LRC circuit portion, such as the one shown in Fig. 1.31, can be calculated by taking into account that

$$e = e_L + e_R + e_C \tag{1.138}$$

The Laplace transform of Eq. (1.138) can be taken by means of the voltages across the inductor, resistor, and capacitor, such that the complex impedance becomes

$$Z(s) = \frac{E(s)}{I(s)} = Ls + R + \frac{1}{Cs} \tag{1.139}$$

It can also be shown that the complex impedance of a parallel LRC circuit is

$$Z(s) = \frac{E(s)}{I(s)} = \frac{1}{1/(Ls) + 1/R + Cs} \tag{1.140}$$

One common configuration where the input and output signals are connected by means of two complex impedances is shown in Fig. 1.39. The transfer function for this system is

$$\frac{E_o(s)}{E_i(s)} = \frac{Z_o(s)}{Z_i(s) + Z_o(s)} \tag{1.141}$$

Example: A direct MEMS application of the mechanical-electrical analogy is found in the field of resonator filters. Figure 1.40 illustrates a filter with n stages, each stage consisting of a microresonator. The input to the system is either a voltage or a current, and likewise, the system output is either a voltage or a current, as mentioned by Lin, Howe, and Pisano,[21] for instance. The

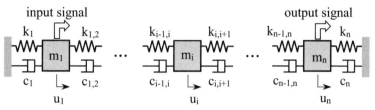

Figure 1.40 Filter with n mechanical resonator stages.

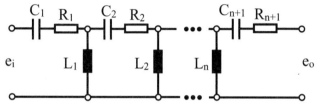

Figure 1.41 Electrical analog of the mechanical system of Fig. 1.40.

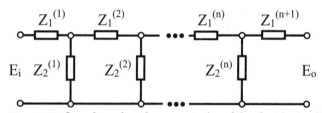

Figure 1.42 Impedance-based representation of the electric circuit of Fig. 1.41.

resonant interaction between the stages modifies (filters) the input signal to a desired output value, which can be tailored through design of the mechanical properties (masses and stiffnesses) of the resonator stages.

The electrical system which is analogous to the mechanical system of Fig. 1.40 is shown in Fig. 1.41. Likewise, the electrical system of Fig. 1.41 can be represented by means of the complex impedances shown in Fig. 1.42. It can be shown that the overall transfer function for the system sketched in Fig. 1.42 is:

$$
\mathrm{TF} = \frac{E_o(s)}{E_i(s)} \times \frac{L_1 s}{R_1 + 1/(C_1 s) + L_1 s} \times \frac{L_2 s}{R_2 + 1/(C_2 s) + L_2 s} \times \cdots
$$

$$
\times \frac{L_{n-1} s}{R_{n-1} + 1/(C_{n-1} s) + L_{n-1} s} \times \frac{R_{n+1} + 1/(C_{n+1} s) + L_n s}{R_n + 1/(C_n s) + L_n s}
$$

(1.142)

References

1. S. Timoshenko, *Vibration Problems in Engineering*, D. Van Nostrand Company, New York, 1928.

2. W. T. Thomson, *Theory of Vibration with Applications*, 2d ed., Prentice-Hall, Englewood Cliffs, NJ, 1981.

3. S. S. Rao, *Mechanical Vibrations*, 2d ed., Addison-Wesley, New York, 1990.

4. K. Ogata, *System Dynamics*, 4th ed., Prentice-Hall, New York, 2004.

5. J. B. Starr, Squeeze-film damping in solid-state accelerometers, *Technical Digest*, IEEE Solid State Sensor and Actuator Workshop, 1990, pp. 44–47.

6. M. Andrews, I. Harris, and G. Turner, A comparison of squeeze-film theory with measurements on a microstructure, *Sensors & Actuators*, **36**, 1993, pp. 79–87.

7. T. Veijola, H. Kuisma, and J. Lahdenpera, Model for gas film damping in a silicon accelerometer, *International Conference on Solid-State Sensors and Actuators*, **4**, 1997, pp. 1097–1100.

8. J. J. Blech, On isothermal squeeze films, *Journal of Lubrication Technology*, **105**, 1983, pp. 615–620.

9. W. C. Tang, T.-C. Nguyen, and R.T. Howe, Laterally-driven polysilicon resonant structures, *Proceedings of IEEE Micro Electro Mechanical Systems Conference*, Oiso, Japan, 1989, pp. 53–59.

10. Y.-H. Cho, A. P. Pisano, and R. T. Howe, Viscous damping model for laterally oscillating microstructures, *Journal of Microelectro-mechanical Systems*, 3(2), 1994, pp. 81–87.

11. X. Zhang, and W. C. Tang, Viscous air damping in laterally-driven microresonators, *Sensor and Actuators*, **20**(1-2), 1989, pp. 25–32.

12. N. Yazdi, F. Ayazi, and K. Najafi, Micromachined inertial sensors, *Proceedings of the IEEE*, **86**(8), 1998, pp. 1640–1659.

13. T. W. Roszhart, The effect of thermoelastic internal friction on the *Q* of micromachined silicon resonators, *Technical Digest on Solid-State Sensor and Actuator Workshop*, 1990, pp. 13–16.

14. R. Lifshitz, and M. L. Roukes, Thermoelastic damping in micro- and nanomechanical systems, *Physical Review B*, **61**(8), 2000, pp. 5600–5609.

15. K. Y. Yasumura, T. D. Stowe, E. M. Chow, T. Pfafman, T. W. Kenny, B. C. Stipe, and D. Rugar, Quality factors in micron- and submicron-thick cantilevers, *Journal of Microelectromechanical Systems*, 9(1), 2000, pp. 117–125.

16. H. Hosaka, K. Itao, and S. Kuroda, Damping characteristics of beam-shaped micro-oscillators, *Sensors and Actuators A*, **49**, 1995, pp. 87–95.

17. D. A. Czaplewski, J. P. Sullivan, T. A. Friedmann, D. W. Carr, B. E. N. Keeler, and J. R. Wendt, Mechanical dissipation in tetrahedral

amorphous carbon, *Journal of Applied Physics*, **97**(2), 2005, pp. 023517-1–023517-10.

18. A. M. Freudenthal, *Introduction to the Mechanics of Solids*, Wiley, New York, 1966.

19. A. N. Cleland, and M. L. Rourkes, Noise processes in nanomechanical resonators, *Journal of Applied Physics*, **92**(5), 2002, pp. 2758–2769.

20. R. Mazet, *Mécanique Vibratoire*, Dunod, Paris, 1966.

21. L. Lin, R. T. Howe, and A. P. Pisano, Microelectromechanical filters for signal processing, *Journal of Microelectromechanical Systems*, **7**(3), 1998, pp. 286–294.

2

Basic Members: Lumped- and Distributed-Parameter Modeling and Design

2.1 Introduction

In this chapter we study the methods of determining the resonant frequencies of basic micromembers such as one-dimensional or two-dimensional ones by using the lumped-parameter modeling and the distributed-parameter technique. Microhinges, microcantilevers, and microbridges, in their most common configurations, can be modeled as line elements of either constant or variable cross section. More specifically, microhinges and microcantilevers can be characterized in terms of their resonant behavior by means of lumped-parameter elastic and inertia properties defined about 6 degrees of freedom that are associated to the free endpoint, namely, three translations (u_x, u_y, and u_z) and three rotations (θ_x, θ_y, and θ_z), as suggested in Fig. 2.1. These degrees of freedom are physical deformations (either linear or rotary) of the member itself and are produced by actuation or interaction with supports and/or adjacent members through bending, torsion, and/or axial loading. As shown in the following, these degrees of freedom are related to the corresponding loads (the forces F_x, F_y, F_z and the moments M_x, M_y, M_z) in the static domain by means of stiffnesses or by means of compliances.

The stiffness and mass can be lumped at the free end of a fixed-free (microcantilever) member, such as the one in Fig. 2.1. The lumped stiffness k_i will be evaluated as the ratio of an applied force/moment

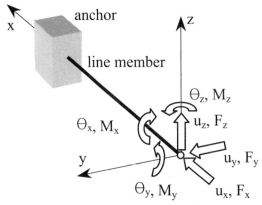

Figure 2.1 Fixed-free line member with 6 degrees of freedom as a schematic representation for microhinges and microcantilevers.

and the corresponding deflection/rotation at the free endpoint and is determined by methods pertaining to the mechanics and strength of materials. The lumped inertia m_i will be calculated by applying Rayleigh's approximate method, which produces an equivalent (or effective) inertia fraction. The resonant frequency about a specific degree of freedom will be calculated as

$$\omega_i = \sqrt{\frac{k_i}{m_i}} \tag{2.1}$$

This chapter focuses on determining the lumped-parameter undamped resonant properties (stiffness, effective inertia fraction, and frequency) of straight-line members constructed by using a single geometric curve, which can be a straight line, a circle, or an ellipse, in order to define a specific configuration. These basic designs either can be used as stand-alone mechanisms (the case of microcantilevers, for instance, which is studied in this chapter) or can be combined and incorporated into more complex shapes of microcantilevers, microhinges, or microbridges, as detailed in Chaps. 3 and 4.

The distributed-parameter approach section, which concludes this chapter, presents exact and approximate methods for directly calculating the value of a relevant resonant frequency, without having to resort to separate evaluations of the corresponding stiffness and mass fractions, as in lumped-parameter modeling. Line members which are subject to axial loads (stresses) can be characterized at resonance by this approach, as well as rings, thin plates, and membranes.

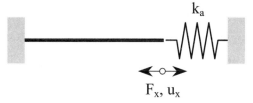

$$k_a$$

$$F_x, u_x$$

Figure 2.2   Lumped-parameter axial stiffness model.

2.2   Lumped-Parameter Modeling and Design

The largest segment of this chapter is dedicated to presenting the lumped-parameter technique applied to line members which can model microcantilevers, microhinges, or microbridges. This method furnishes fractions for both stiffness and inertia which can be used together to calculate the resonant frequency corresponding to a relevant degree of freedom.

2.2.1   Lumped-parameter stiffnesses and compliances

To find a lumped-parameter stiffness k_i which is associated with bending, axial loading, or torsion means to determine a relationship of the type

$$L_i = k_i d_i \qquad (2.2)$$

where the load L_i is either a force (F_x, F_y, or F_z) or a moment (M_x, M_y, or M_z), whereas the displacement/deformation d_i is either a linear quantity (u_x, u_y, or u_z) or an angular one (θ_x, θ_y, or θ_z), as suggested in Fig. 2.1. The illustration of the generic Eq. (2.2) in the case of axial loading and deformation is sketched in Fig. 2.2, where a force applied at the free end about the longitudinal (x) direction of the fixed-free member produces an elastic deformation about the same direction at that point. This elastic interaction can be modeled by a linear spring of stiffness k_a.

Similarly, a moment that is applied about the longitudinal (x) axis at the free end will generate an angular deformation at that point, and this interaction can be modeled by a torsional spring of stiffness k_t, as illustrated in Fig. 2.3.

In bending (about the y axis, for instance), three types of stiffnesses can be identified. The direct linear stiffness k_l, which is modeled by means of a linear spring, as shown in Fig. 2.4, is based on a force-deflection relationship and is similar to the previously defined axial stiffness. The direct rotary stiffness k_r is illustrated in Fig. 2.5, and it

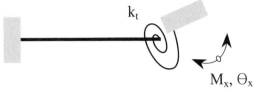

Figure 2.3 Lumped-parameter torsional stiffness model.

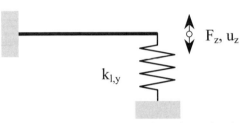

Figure 2.4 Lumped-parameter direct linear bending stiffness for a microcantilever.

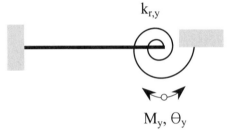

Figure 2.5 Lumped-parameter direct rotary bending stiffness for a microcantilever.

indicates a moment-rotation relationship. Its physical representation is a torsional (spiral) spring, similar to the one used to model torsion.

Eventually, the cross stiffness k_c is pictured in Fig. 2.6, which suggests either a moment-deflection relationship or a force-rotation one. The former interaction can be modeled by a moment that is applied to the eccentric in Fig. 2.6 and will act upon the endpoint by deforming (deflecting) the microcantilever linearly.

Lumped-parameter stiffnesses can be determined in two different manners: either by following a direct approach or by first determining the compliances, which are the stiffness inverses (in either strictly algebraic sense—for axial and torsional loading—or in a matrix sense—for bending), as shown in the following.

Direct stiffness approach. The direct approach of determining stiffnesses usually employs energy methods, such as Castigliano's first

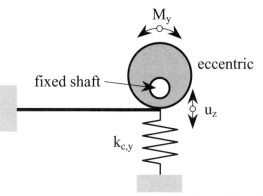

Figure 2.6 Lumped-parameter cross-bending stiffness for a microcantilever.

theorem, which expresses the load L_i that is applied at a point on an elastic body as the partial derivative of the strain energy U evaluated with respect to the corresponding elastic deformation d_i about the load direction at the point of interest, or

$$L_i = \frac{\partial U}{\partial d_i} \tag{2.3}$$

This approach is implemented with ease when the cross section is constant, but is more difficult to utilize for variable-cross-section members. The strain energy corresponding to axial loading can be expressed as

$$U_a = \frac{1}{2E} \int_0^l \frac{F_a^2(x)}{A(x)} = \frac{E}{2} \int_0^l A(x) \left[\frac{du_x(x)}{dx} \right]^2 dx \tag{2.4}$$

where $F_a(x)$ is the axial load, $A(x)$ is the cross-sectional area, $u_x(x)$ is the axial deformation, and E is Young's modulus of elasticity. The differential equation expressing the static equilibrium in axial loading is

$$\frac{d^2[u_x(x)A(x)]}{dx^2} = 0 \tag{2.5}$$

The axial deformation at a generic point on the micromember can be expressed as a function of the axial deformation at the free end and a distribution function $f_a(x)$ as

$$u_x(x) = f_a(x)u_x \tag{2.6}$$

The axial distribution function is determined by considering it as a first-degree polynomial (with two unknown coefficients), which has to satisfy the following boundary conditions that apply to the fixed-free member of Fig. 2.1:

$$u_x(0) = u_x \qquad u_x(l) = 0 \qquad (2.7)$$

It follows that

$$f_a(x) = 1 - \frac{x}{l} \qquad (2.8)$$

The axial stiffness is found by combining Eqs. (2.3) through (2.6) as

$$k_a = E \int_0^l A(x) \left[\frac{df_a(x)}{dx} \right]^2 dx \qquad (2.9)$$

A similar procedure is applied to torsion, where the strain energy is

$$U_t = \frac{1}{2G} \int_0^l \frac{M_t^2(x)}{I_t(x)} = \frac{G}{2} \int_0^l I_t(x) \left[\frac{d\theta_x(x)}{dx} \right]^2 dx \qquad (2.10)$$

In Eq. (2.10), $M_t(x)$ is the torsional moment, $I_t(x)$ is the torsional moment of inertia, $\theta_x(x)$ is the angular deformation, and G is the shear modulus of elasticity. The differential equation expressing the static equilibrium in torsion is

$$\frac{d^2[\theta_x(x)I_t(x)]}{dx^2} = 0 \qquad (2.11)$$

The torsion angle at an abscissa x is expressed in terms of the torsion angle at the free end by means of a distribution function $f_t(x)$ as

$$\theta_x(x) = f_t(x)\theta_x \qquad (2.12)$$

By combining Eqs. (2.3) and (2.10) through (2.12), the torsional stiffness is

$$k_t = G \int_0^l I_t(x) \left[\frac{df_t(x)}{dx} \right]^2 dx \qquad (2.13)$$

It can be shown that the torsion distribution function is identical to the axial one [Eq. (2.8)]. Later in this chapter, inertia fractions are derived that correspond to axial or torsional free vibrations. These inertia fractions are determined in terms of distribution functions that are identical to the one expressed in Eq. (2.8).

The bending presents the characteristic of effect coupling, as deflection or rotation can be produced by both a moment and a force. The strain energy in bending about the y axis is expressed as

$$U_{b,y} = \frac{1}{2E} \int_0^l \frac{M_{b,y}^2(x)}{I_y(x)} \, dx = \frac{E}{2} \int_0^l I_y(x) \left[\frac{d^2 u_z(x)}{dx^2} \right]^2 dx \qquad (2.14)$$

where $M_{b,y}(x)$ is the bending moment about the y axis, $I_y(x)$ is the moment of inertia calculated about the same axis, and $u_z(x)$ is the deflection about the z axis. The differential equation for the bending static equilibrium is

$$\frac{d^4[I_y(x)u_z(x)]}{dx^4} = 0 \qquad (2.15)$$

The deflection at a generic point on the micromember length is expressed in terms of the free end deflection and rotation, as well as two distribution functions: the linear one $f_l(x)$ and the rotary one $f_r(x)$ in the form:

$$u_z(x) = f_l(x)u_z + f_r(x)\theta_y \qquad (2.16)$$

The two distribution functions of Eq. (2.16) are determined as third-degree polynomials (with four unknown coefficients each) by imposing the following boundary conditions:

$$u_z(0) = u_z \quad \theta_y(0) = \theta_y \quad u_z(l) = 0 \quad \theta_y(l) = 0 \qquad (2.17)$$

The bending-related distribution functions are found to be

$$f_l(x) = 1 - \frac{3x^2}{l^2} + \frac{2x^3}{l^3} \qquad f_r(x) = x - \frac{2x^2}{l} + \frac{x^3}{l^2} \qquad (2.18)$$

We show later in this chapter, when deriving inertia fractions, that other distribution functions need to be utilized when quantifying the mass fraction (called the effective mass) that corresponds to bending.

By combining Eqs. (2.3) and (2.14) through (2.16), the three bending-related stiffnesses (direct linear, direct rotary, and cross) are expressed as

$$k_{l,y} = E \int_0^l I_y(x) \left[\frac{d^2 f_l(x)}{dx} \right]^2 dx$$

$$k_{r,y} = E \int_0^l I_y(x) \left[\frac{d^2 f_r(x)}{dx} \right]^2 dx \qquad (2.19)$$

$$k_{c,y} = E \int_0^l I_y(x) \frac{d^2 f_l(x)}{dx} \frac{d^2 f_r(x)}{dx} dx$$

The bending about the other axis, the z axis, can be treated similarly, and the equations that have been presented for the bending about the y axis are valid for the z axis by interchanging of the y and z subscripts in Eqs. (2.14) through (2.19). The stiffness matrix collecting axial, torsion, and two-axis bending effects can be expressed as

$$[K] = \begin{bmatrix} k_{l,y} & k_{c,y} & 0 & 0 & 0 & 0 \\ k_{c,y} & k_{r,y} & 0 & 0 & 0 & 0 \\ 0 & 0 & k_{l,z} & k_{c,z} & 0 & 0 \\ 0 & 0 & k_{c,z} & k_{r,z} & 0 & 0 \\ 0 & 0 & 0 & 0 & k_a & 0 \\ 0 & 0 & 0 & 0 & 0 & k_t \end{bmatrix} \qquad (2.20)$$

and a load deformation equation can be expressed in matrix form as

$$\{L\} = [K]\{d\} \qquad (2.21)$$

where the load vector is

$$\{L\} = \left\{ F_z \, M_y \, F_y \, M_z \, F_x \, M_x \right\}^T \qquad (2.22)$$

and the deformation/displacement vector is

$$\{d\} = \left\{ u_z \, \theta_y \, u_y \, \theta_z \, u_x \, \theta_x \right\}^T \qquad (2.23)$$

Compliance approach. The second approach to finding the relevant stiffnesses of fixed-free straight line members addresses the compliance (or flexibility) method, which utilizes Castigliano's second (or displacement) theorem to find the relationship between an elastic deformation at a point along a specified direction and the corresponding load, by using the partial derivative of the strain energy, namely,

$$d_i = \frac{\partial U}{\partial L_i} \tag{2.24}$$

In axial loading, the corresponding compliance is

$$C_a = \frac{1}{E} \int_0^l \frac{dx}{A(x)} \tag{2.25}$$

Similarly, the torsion-related compliance is calculated as

$$C_t = \frac{1}{G} \int_0^l \frac{dx}{I_t(x)} \tag{2.26}$$

For bending about the y axis, the direct linear, direct rotary, and cross compliances are evaluated as

$$C_{l,y} = \frac{1}{E} \int_0^l \frac{x^2 \, dx}{I_y(x)} \qquad C_{r,y} = \frac{1}{E} \int_0^l \frac{dx}{I_y(x)} \qquad C_{c,y} = \frac{1}{E} \int_0^l \frac{x \, dx}{I_y(x)} \tag{2.27}$$

The first of Eqs. 3. (2.27) is valid for microcantilevers that are relatively long (where the length is at least 5 times larger than the largest cross-sectional dimension, which is generally the width—see Young and Budynas,[1] for instance) and where the Euler-Bernoulli assumptions and model do apply. For relatively short configurations, shearing deformations add to the ones normally produced by bending such that the linear compliance is expressed according to the Timoshenko model (which is presented in greater detail a bit later in this chapter) as follows (see Lobontiu[2] for more details):

$$C_{l,y}^{sh} = C_{l,y} + \kappa \frac{E}{G} C_a \tag{2.28}$$

where κ is a constant accounting for the micromember's cross-sectional shape.

The compliances corresponding to bending about the z axis are expressed by using the y-z subscript interchange in Eqs. (2.27) and (2.28), which define the bending compliances corresponding to the y axis. In the end, the compliances corresponding to axial, torsion, and two-axis bending can be arranged into the following compliance matrix:

$$[C] = \begin{bmatrix} C_{l,y} & C_{c,y} & 0 & 0 & 0 & 0 \\ C_{c,y} & C_{r,y} & 0 & 0 & 0 & 0 \\ 0 & 0 & C_{l,z} & C_{c,z} & 0 & 0 \\ 0 & 0 & C_{c,z} & C_{r,z} & 0 & 0 \\ 0 & 0 & 0 & 0 & C_a & 0 \\ 0 & 0 & 0 & 0 & 0 & C_t \end{bmatrix} \qquad (2.29)$$

such that the deformation/displacement vector and the load vector corresponding to the free end of the micromember shown in Fig. 2.1 are connected as

$$\{d\} = [C]\{L\} \qquad (2.30)$$

It is obvious that the compliance and stiffness matrices are inverses of each other, namely,

$$[C] = [K]^{-1} \qquad (2.31)$$

A few remarks are in order here. While the axial- and torsion-related stiffnesses are the algebraic inverses (reciprocals) of their corresponding compliances, namely,

$$k_a = \frac{1}{C_a} \qquad k_t = \frac{1}{C_t} \qquad (2.32)$$

the same is not true for the bending-related stiffness-compliance pairs, due to the coupled effects between deflections/rotations and forces/moments. In other words, it appears, at least from a formal point of view, that

$$k_l \neq \frac{1}{C_l} \qquad k_r \neq \frac{1}{C_r} \qquad k_c \neq \frac{1}{C_c} \qquad (2.33)$$

In cases where forces/moments need to be calculated in terms of known deformations (such as when these are available experimentally), the stiffnesses of Eqs. (2.33) have to be calculated either by applying the direct approach or by inverting the compliance matrix—a more detailed explanation of this aspect is given in Lobontiu and Garcia[3] and Lobontiu et al.[4] It can be shown, however, that the bending-related stiffness, which corresponds to the direct linear effects and which has to be employed in resonant frequency calculations, is the algebraic inverse of the corresponding compliance. Therefore, in this book, the

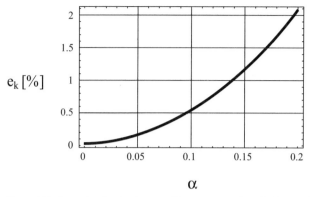

Figure 2.7   Long-to-short microcantilever comparison in terms of the bending stiffness.

bending stiffness is calculated in this way, as it is known that evaluating compliances first is simpler, especially for variable-cross-section members, compared to directly computing stiffnesses.

Example: Compare the bending stiffness of a long microcantilever to that of a short one. The constant rectangular cross section is defined by the width w and the thickness t (width w is the largest dimension and is parallel to the y axis while t is parallel to the z axis in Fig. 2.1), and the member's length is l.

By using the nondimensional parameter,

$$\alpha = \frac{t}{l} \tag{2.34}$$

the following relative error function can be defined:

$$e_k = \frac{k_{l,y} - k_{l,y}^{sh}}{k_{l,y}} = 52.083\alpha^2 \tag{2.35}$$

which is the relative error between the bending stiffness of a long microcantilever and that of a short one. The stiffness $k_{l,y}$ has been calculated as the inverse of the compliance defined in Eqs. (2.27), whereas the stiffness $k_{l,y}^{sh}$ is the inverse of the shearing linear compliance of Eq. (2.28). Figure 2.7 is the two-dimensional plot of this error function. As the figure shows it, the errors are larger than 1 percent only for cases where the thickness t is in excess of 10 percent of the length.

2.2.2   Lumped-parameter inertia properties

To enable resonant frequency calculations pertaining to axial, torsional, or bending loading for microhinges and microcantilevers by means of equations of the type (2.1), lumped-parameter inertia properties need to be established, in addition to the lumped-parameter stiffnesses already discussed. As mentioned previously, only the direct linear bending stiffness (as the algebraic inverse of the corresponding compliance) is important in terms of bending, in addition to the axial- and torsion-related stiffness. As a consequence, inertia properties will be sought that are associated with axial, torsional, and direct linear bending free vibrations.

The procedure enabling calculation of these inertia fractions that are going to be located at the free end of a fixed-free micromember follows the Rayleigh principle. This principle, as mentioned by Timoshenko[5] or Thomson,[6] states that the velocity distribution over a freely vibrating member is identical to the static deformation distribution of the same member. This statement permits calculation of the so-called equivalent (or effective) inertia fraction [either translatory (therefore a mass fraction) or rotary (therefore a mechanical moment of inertia)] by equating the kinetic energy of the real distributed-parameter system to that of the lumped-parameter inertia particle which is located at the free end of the micromember.

Axial vibrations   In axial loading, this principle translates to the following relationships between the generic point deformation/velocity and the free endpoint deformation/velocity, by means of the axial distribution function $f_a(x)$:

$$u_x(x) = f_a(x)u_x \qquad \frac{du_x(x)}{dt} = f_a(x)\frac{du_x}{dt} \qquad (2.36)$$

The kinetic energy of the distributed-parameter fixed-free microrod of Fig. 2.1 undergoing free vibrations is

$$T_a = \frac{1}{2}\rho\int_0^l A(x)\left[\frac{du_x(x)}{dt}\right]^2 dx \qquad (2.37)$$

The kinetic energy of the effective mass is

$$T_{a,e} = \frac{1}{2}m_{a,e}\left(\frac{du_x}{dt}\right)^2 \qquad (2.38)$$

The distribution function corresponding to free axial vibrations of a microcantilever is explicitly formulated later in this chapter when we show that constant- and variable-cross-section configurations have different distribution functions.

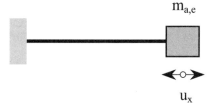

$m_{a,e}$

u_x

Figure 2.8 Effective mass corresponding to free axial vibrations of a fixed-free microrod.

By equating Eqs. (2.37) and (2.38), the effective mass corresponding to axial free vibrations is

$$m_{a,e} = \rho \int_0^l f_a^2(x)A(x)\,dx \qquad (2.39)$$

and is physically represented by the body mass which is placed at the cantilever's free end, as shown in Fig. 2.8.

Torsional vibrations. In torsion, the effective inertia which corresponds to the free vibrations is sketched in Fig. 2.9. An approach similar to the one applied to axial free vibrations is used here, according to which the kinetic energy of the distributed-parameter system is

$$T_t = \frac{1}{2}\rho \int_0^l I_t(x)\left[\frac{d\theta_x(x)}{dt}\right]^2 dx \qquad (2.40)$$

The kinetic energy of the effective, lumped-parameter system is

$$T_{t,e} = \frac{1}{2}J_t\left(\frac{d\theta_x}{dt}\right)^2 \qquad (2.41)$$

Based on Rayleigh's assumption, the following relationships can be written:

$$\theta_x(x) = f_t(x)\theta_x \qquad \frac{d\theta_x(x)}{dt} = f_t(x)\frac{d\theta_x}{dt} \qquad (2.42)$$

By equating Eqs. (2.40) and (2.41), the lumped-parameter effective inertia fraction corresponding to free torsional vibrations is found to be

$$J_{t,e} = \rho \int_0^l f_t^2(x)I_t(x)\,dx \qquad (2.43)$$

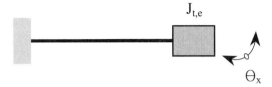

Figure 2.9 Effective inertia corresponding to free torsional vibrations of a fixed-free microbar.

Figure 2.10 Effective mass corresponding to free bending vibrations of a fixed-free microbeam.

Again, the distribution function in torsion is determined later in this chapter for both constant- and variable-cross-section members.

Bending vibrations. The linear oscillatory motion that occurs during the free bending vibrations of a fixed-free microbeam can be modeled in terms of lumped-parameter inertia by an effective mass located at the member's free end, as shown in Fig. 2.10.

An approach similar to the one taken for free axial vibrations produces the following lumped mass:

$$m_{b,e} = \rho \int_0^l f_b^2(x) A(x)\, dx \qquad (2.44)$$

As mentioned previously for free axial and torsional vibrations, the distribution function corresponding to bending free vibrations is explicitly derived for both constant- and variable-cross-section configurations later in this chapter.

2.2.3 Constant-cross-section members

Microcantilevers and microhinges of constant cross section (generally rectangular) are first analyzed, and the first resonant frequencies are calculated. It can be shown that for thin fixed-free (as well as for fixed-fixed) components the first resonant frequency corresponds to bending, and therefore both the lumped-parameter stiffness and inertia are determined by studying the bending about the sensitive axis (the y axis in Fig. 2.1).

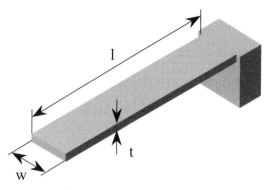

Figure 2.11   Constant rectangular cross-section microcantilever.

As mentioned previously, microcantilevers are fixed-free members, which can resonantly vibrate in bending, torsion, and/or axially. The sketch of a constant rectangular cross-section microcantilever is shown in Fig. 2.11. The axial and torsional resonant frequencies will be determined first, followed by the first bending resonant frequency.

The aim here is to equivalently transform the distributed-parameter microcantilever into a lumped-parameter system, which will enable formulation of the relevant stiffness and mass such that a particular natural frequency be calculated by means of Eq. (2.1).

Axial vibrations.   The particular situation of axial vibrations is pictured in Fig. 2.12, which shows the original, distributed-parameter system (Fig. 2.12a) and the equivalent lumped-parameter one (Fig. 2.12b). It is well known that the lumped-parameter stiffness at the end of the axially vibrating rod is

$$k_{e,a} = \frac{EA}{l} \qquad (2.45)$$

This equation is obtained by both the stiffness and the compliance approaches, as it can be easily verified by applying the two procedures just presented in this chapter.

The equivalent inertia fraction which has to be placed at the free extremity of the microrod sketched in Fig. 2.12b is calculated by means of Rayleigh's approximate method, as shown previously, according to which the distribution of the velocity field of a vibrating component is identical to the displacement (axial deflection here) distribution of the same component. By equating the kinetic energy of the real, distributed-parameter system to the kinetic energy of the equivalent, lumped-parameter system, an equivalent (or effective) mass is produced.

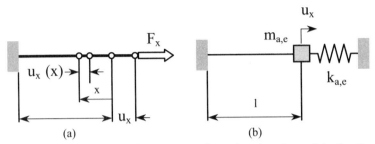

Figure 2.12 (*a*) Distributed-parameter microrod undergoing axial vibration; (*b*) equivalent lumped-parameter mass-spring system.

The deflection at an abscissa x measured from the free end of the microrod of Fig. 2.12*a*, $u_x(x)$, is related to the free end's deflection u_x by a distribution function $f_a(x)$ in the form:

$$u_x(x) = f_a(x)u_x(0) = f_a(x)u_x \qquad (2.46)$$

where the distribution function is the one determined in Eq. (2.8)

The kinetic energy of the distributed-parameter microrod corresponding to axial vibration about the x axis is

$$T_a = \frac{\rho A}{2}\int_0^l \dot{u}_x(x)^2 dx = \frac{\rho A \dot{u}_x^2}{2}\int_0^l f_a^2(x)dx \qquad (2.47)$$

The kinetic energy of a mass $m_{a,e}$ which is placed at the free end 1 is

$$T_{a,e} = \frac{m_{a,e}\dot{u}_x^2}{2} \qquad (2.48)$$

By equating Eq. (2.47) to Eq. (2.48), the equivalent mass is found to be

$$m_{a,e} = \frac{m}{3} \qquad (2.49)$$

where m is the total mass of the microrod. By combining Eqs. (2.1), (2.45), and (2.49), the resonant frequency becomes

$$\omega_{a,e} = 1.732\sqrt{\frac{EA}{ml}} \qquad (2.50)$$

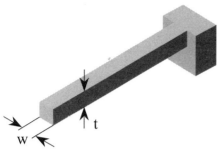

Figure 2.13 Constant rectangular cross-section microbar.

Torsional vibrations. A similar path can be followed to determine the resonant frequency of a fixed-free bar undergoing torsional vibrations. The lumped stiffness at the free tip of the bar corresponding to torsional deformation is

$$k_{t,e} = \frac{GI_t}{l} \qquad (2.51)$$

The torsional moment of inertia of a constant rectangular cross-section with $t \ll w$ (this configuration is referred to as very thin) can be expressed (see Boresi, Schmidt, and Sidebottom,[7] for instance) as

$$I_t = \frac{wt^3}{3} \qquad (2.52)$$

There are designs where the cross section is relatively thick, with the dimensions w and t being comparable, as sketched in Fig. 2.13.

For such designs, as shown in Young and Budynas[1] or Lobontiu,[2] the torsional moment of inertia can be approximated to

$$I_t = wt^3 \left(0.33 - \frac{0.21t}{w} \right) \qquad (2.53)$$

Circular cross-section microbars, such as those constructed of carbon nanotubes, as shown in Fig. 2.14, are defined by the torsional moment of inertia

$$I_t = \frac{\pi d^4}{32} \qquad (2.54)$$

where d is the diameter.

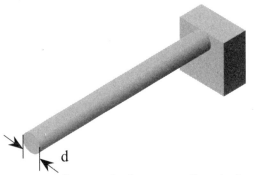

Figure 2.14 Constant circular cross-section microbar.

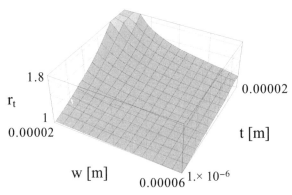

Figure 2.15 Very thin versus thin rectangular cross-section bars in terms of the moment of inertia.

Example: Compare the torsional moment of inertia of a very thin constant rectangular cross-section bar ($t \ll w$) to that of a thin constant rectangular cross-section bar ($t < w$).

If the ratio of the two moments of inertia—the one of Eq. (2.52) to the one provided in Eq. (2.53)—is analyzed, the plot of Fig. 2.15 can be drawn, where r_t denotes the previously mentioned moment of inertia ratio. As Fig. 2.15 shows, the predictions by the very thin model are always higher than those produced by the thin model (the ratio r_t is larger than 1). For small values of thickness t, the two models yield comparable results; but when the thickness t approaches the magnitude of the width w, the very thin model's predictions can be 1.8 higher than those of the thin model. Care should therefore be exercised to correctly select the appropriate model as a function of the thickness-to-width ratio and the amount of error that is considered tolerable in adopting one model over the other.

In this sense, Fig. 2.16 illustrates the relative errors between the torsional moments of inertia given by the very thin theory—Eq. (2.52)—and by the thin theory—Eq. (2.53). It can be observed that the errors between the two models

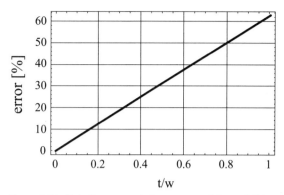

Figure 2.16   Relative errors between very thin and thin cross-section torsional moments of inertia.

are less than 5 percent for thickness-to-width ratios less than 0.1, and therefore it might appear reasonable to use the very thin model up to this geometry limitation. Anyway, as previously noted, when the thickness-to-width ratio approaches 1, the errors between the two predictions can be as high as 60 percent, and this is shown in the same Fig. 2.16.

An equivalent mass which is placed at the free end of the microrod and undergoes rotation about the x axis is found again by equating the kinetic energy of the real, distributed-parameter microbar to the kinetic energy of the equivalent (effective) mass; its equation is

$$J_{t,e} = \frac{J_t}{3} \tag{2.55}$$

where J_t is the mechanical moment of inertia of the rectangular cross-section microbar and is given by

$$J_t = \frac{m(w^2 + t^2)}{12} \tag{2.56}$$

For a circular cross section, J_t is

$$J_t = \frac{md^2}{8} \tag{2.57}$$

as given by Beer and Johnston.[8]

The torsional resonant frequency can be expressed as

$$\omega_{t,e} = 1.732 \sqrt{\frac{GI_t}{J_t l}} \tag{2.58}$$

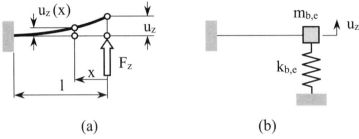

(a)   (b)

Figure 2.17 (*a*) Distributed-parameter microcantilever; (*b*) equivalent lumped-parameter microcantilever.

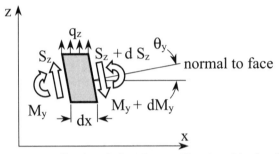

normal to face

Figure 2.18 Portion of a long beam with external load and internal reactions.

Bending vibrations. The bending vibration of a microcantilever is illustrated in Fig. 2.17. The original distributed-parameter member is pictured in Fig. 2.17*a* whereas the corresponding lumped-parameter (mass-spring) system is sketched in Fig. 2.17*b*. Two possibilities are studied here, namely, the long configuration (following the Euler-Bernoulli beam model) and the relatively short configuration, which is described by the Timoshenko model.

Long microcantilevers. In the case of relatively long beam configurations (where the length is at least 5 times larger than the largest cross-sectional dimension), the tangent to the neutral axis is perpendicular to the face, as sketched in Fig. 2.18, which pictures a portion removed from a deformed (bent) beam.

The static equilibrium can be analyzed for this segment under the action of the external distributed load q_z, shearing forces S_z, and bending moments M_y. As known from the mechanics of materials, the bending moment can be expressed as

$$M_y(x) = EI_y \frac{d^2 u_z(x)}{dx^2} \tag{2.59}$$

By integrating this equation twice and by applying the adequate boundary conditions of the beam shown in Fig. 2.17a, which are zero slope and zero deflection at the fixed end

$$\frac{du_z(x)}{dx}\bigg|_{x=l} = 0 \quad u_z(l) = 0 \tag{2.60}$$

the deflection equation can be found. It is known from the mechanics of materials that the stiffness at the point is the ratio of a force that is applied at that point about the z axis to the corresponding deflection, and it is calculated as

$$k_{b,e} = \frac{3EI_y}{l^3} \tag{2.61}$$

This is actually the inverse of the compliance $C_{l,y}$ of Eq. (2.27), which has been calculated by means of the compliance approach. A different value would be obtained for $k_{b,e}$ if it were calculated by means of the stiffness approach, as noted in that subsection of this chapter. That value, however, is not the one needed for resonant frequency calculations, and therefore Eq. (2.61) is the one applicable for such purposes, as demonstrated shortly.

The equivalent inertia fraction which has to be placed at the free extremity of the microcantilever sketched in Fig. 2.17b is calculated again by means of Rayleigh's approximate method, according to which the distribution of the velocity field of a bending vibrating component is identical to the displacement (deflection here) distribution of the same component. By equating the kinetic energy of the real, distributed-parameter system to the kinetic energy of the equivalent, lumped-parameter system, an equivalent (or effective) mass is produced. The deflection at an abscissa x measured from the free end of the microcantilever of Fig. 2.17a, $u_z(x)$, is related to the free end's deflection u_z by means of a distribution function $f_b(x)$ in the form:

$$u_z(x) = f_b(x)u_z(0) = f_b(x)u_z \tag{2.62}$$

where the distribution function is (see Lobontiu,[2] for instance)

$$f_b(x) = 1 - \frac{3x}{2l} + \frac{x^3}{2l^3} \tag{2.63}$$

This distribution function can simply be determined by expressing the deflection at a generic point, of abscissa x, as a function of the deflection at the free end of a cantilever beam. It can be seen that the deflection distribution law and function are different from those of Eqs. (2.16) and (2.18), which dealt with the stiffness approach.

The kinetic energy of the distributed-parameter microcantilever corresponding to bending about the z axis is

$$T_b = \frac{\rho A}{2} \int_0^l \dot{u}_z^2(x)\, dx = \frac{\rho A \dot{u}_z^2}{2} \int_0^l f_b^2(x)\, dx \tag{2.64}$$

The kinetic energy of a mass $m_{b,e}$ which is placed at the free end of the microcantilever is simply

$$T_{b,e} = \frac{m_{b,e} \dot{u}_z^2}{2} \tag{2.65}$$

By equating Eqs. (2.64) and (2.65), the equivalent mass $m_{b,e}$ is found to be

$$m_{b,e} = \frac{33}{140}\, m \tag{2.66}$$

where m is the total mass of the microcantilever. By combining Eqs. (2.1), (2.61), and (2.66), the bending resonant frequency of the microcantilever becomes

$$\omega_{b,e} = 3.567\sqrt{\frac{EI_y}{ml^3}} \tag{2.67}$$

As will be demonstrated in Chap. 5, the exact value of the bending resonant frequency can be calculated by integrating a differential equation, and its value is

$$\omega_b = 3.52\sqrt{\frac{EI_z}{ml^3}} \tag{2.68}$$

In other words, the relative error in the first resonant frequency calculated by the approximate Eq. (2.67) versus the exact Eq. (2.68) is 1.33 percent.

Example: Compare the axial, torsional, and bending resonant frequencies of a long microcantilever having a constant and very thin rectangular cross section ($t \ll w$).

If we use the following nondimensional parameters

$$\alpha = \frac{t}{l} \qquad \beta = \frac{w}{l} \tag{2.69}$$

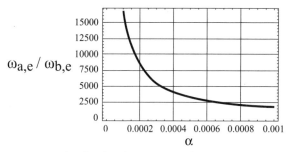

Figure 2.19 Axial-to-bending resonant frequency.

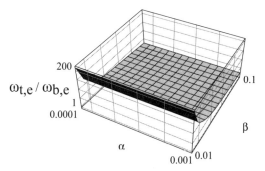

Figure 2.20 Torsional-to-bending resonant frequency ratio.

and take into consideration the definitions of the cross-sectional area A, moment of inertia I_y, torsional moment of inertia I_t, and mechanical torsional moment of inertia J_t, as well as the relationship between the longitudinal and shear elastic moduli, which is

$$G = \frac{E}{2(1+\mu)} \tag{2.70}$$

We obtain the following ratios for Poisson ratio $\mu = 0.25$ corresponding to polysilicon:

$$\frac{\omega_{a,e}}{\omega_{b,e}} = \frac{1.682}{\alpha} \qquad \frac{\omega_{t,e}}{\omega_{b,e}} = \frac{2.128}{\sqrt{\alpha^2 + \beta^2}} \qquad \frac{\omega_{a,e}}{\omega_{t,e}} = 0.79\sqrt{1 + \frac{\beta^2}{\alpha^2}} \tag{2.71}$$

The last of Eqs. (2.71) is a combination of the first two. Figures 2.19, 2.20, and 2.21 are plots of the resonant frequency ratios defined and formulated in Eqs. (2.71). As Figs. 2.19 and 2.20 do suggest, the bending resonant frequency of a long, thin, constant rectangular cross-section fixed-free member is always smaller than both the axial and the torsional resonant frequencies by factors that can be as high as 200 in the case of torsion (Fig. 2.20) and 15,000 in the case of axial free vibrations, as shown in Fig. 2.19. The axial

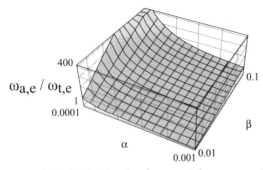

Figure 2.21 Axial-to-torsional resonant frequency ratio.

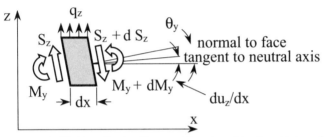

Figure 2.22 Portion of a short beam with external load and internal reactions.

resonant frequency is larger than the torsional one, by a factor of up to 400, as pictured in Fig. 2.21.

Short microcantilevers. For relatively short beams (where the length is less than 5 times the largest cross-sectional dimension, as mentioned previously), the tangent to the neutral axis is no longer perpendicular to the face, as indicated in Fig. 2.22.

This is due to the fact that shearing effects become important for short beams, and they produce an additional angular deformation $\theta_y - du_z/dx$, which is shown in the same figure. The equations describing the combined effects of bending and shearing, according to the Timoshenko model, are

$$M_y(x) = EI_y \frac{d\theta_y(x)}{dx}$$

$$\theta_y(x) - \frac{du_z(x)}{dx} = \frac{\kappa S_z(x)}{AG}$$

(2.72)

where κ is the coefficient accounting for the cross-sectional shape, which is equal to $\frac{5}{6}$ for a rectangular shape. By taking into account that

$$M_y(x) = F_z x \qquad (2.73)$$

for a beam loaded with a tip force F_z, we can find the angle $\theta_y(x)$ by integrating the first of Eqs. (2.72), considering that

$$\theta_y(l) = 0 \qquad (2.74)$$

and its equation is

$$\theta_y(x) = \frac{F_z(l^2 - x^2)}{2EI_y} \qquad (2.75)$$

By substituting $\theta_y(x)$ of Eq. (2.75) into the second of Eqs. (2.72) and by carrying out the necessary integration with the boundary condition

$$u_z(l) = 0 \qquad (2.76)$$

the deflection $u_z(x)$ can be written into the generic form of Eq. (2.62) where the distribution function is

$$f_b^{sh}(x) = (l - x)\frac{GA(2l + x)(l - x) + 6\kappa EI_y}{2l(GAl^2 + 3\kappa EI_y)} \qquad (2.77)$$

and the tip deflection is

$$u_z = u_z(0) = F_z l\left(\frac{l^2}{3EI_y} + \frac{\kappa}{GA}\right) \qquad (2.78)$$

Equation (2.78) enables us to find the shearing-dependent stiffness as

$$k_{b,e}^{sh} = \frac{F_z}{u_z} = \frac{3EGI_yA}{l(GAl^2 + 3\kappa EI_y)} \qquad (2.79)$$

The effective mass which is placed at the free extremity is determined by following Rayleigh's procedure, which has been outlined previously, and by using the distribution function of Eq. (2.77). Its equation is

$$m_{b,e}^{sh} = \frac{3(140\kappa^2 E^2 I_y^2 + 77\kappa EGI_yAl^2 + 11G^2A^2l^4)}{[140(GAl^2 + 3\kappa EI_y)^2]}m \qquad (2.80)$$

where m is the mass of the microcantilever.

The corresponding resonant frequency is found by means of the stiffness given in Eq. (2.79) and the effective mass of Eq. (2.80) as

$$\omega_{b,e}^{sh} = 11.832\sqrt{\frac{EGI_yA(GAl^2 + 3\kappa EI_y)}{ml(140\kappa^2E^2I_y^2 + 77\kappa EGI_yAl^2 + 11G^2A^2l^4)}} \qquad (2.81)$$

It can be shown that by neglecting the shear force S_z, the Timoshenko model produces the equivalent stiffness, mass, and resonant frequency that are given by the Euler-Bernoulli model in Eqs. (2.61), (2.66), and (2.67), respectively.

Example: Compare the inertia distribution functions corresponding to long- and short-beam models. Compare the bending-related resonant frequencies of long and short microcantilevers of constant rectangular cross section.

The ratio of the distribution function for a long member to that of a short member can be expressed by Eqs. (2.77) and (2.63) as

$$r_{fb} = \frac{f_b^{sh}}{f_b} = \frac{2[(8 + 5\alpha^3)l^2 - 4lx - 4x^2]}{(8 + 5\alpha^3)(l - x)(2l + x)} \qquad (2.82)$$

where $\qquad\qquad\qquad\qquad t = \alpha l \qquad\qquad\qquad\qquad (2.83)$

The usual relationship between the longitudinal and transverse elasticity moduli [Eq. (2.70)] has been considered here with a Poisson's ratio of 0.25 (the approximate value for polysilicon) and a shear factor of $\frac{5}{6}$, as indicated in Young and Budynas,[1] for instance. Obviously, $\alpha \ll 1$ because, for thin microcantilevers, $t \ll 1$. It is interesting to analyze the ratio of Eq. (2.82) as a function of α. This can be done in a graphical manner, and Fig. 2.23 illustrates this ratio plotted in terms of α and for $x = 0.99l$. It can be seen that for very small values of α (therefore very thin members), the effects of shear are small (the ratio is slightly larger than 1) and often can be neglected.

When we compare the bending resonant frequencies of long and short microcantilevers of constant rectangular cross section, we use Eqs. (2.67) and (2.81), and Fig. 2.24 is the two-dimensional plot of the long-to-short bending resonant frequency ratio.

For very small values of the parameter α, the resonant frequency provided by the long-beam model is slightly smaller than the one produced by the short-beam model. With α increasing, the roles reverse and the predictions by the long-beam model exceed those provided by the short-beam model. The differences between the two models are, however, small, as Fig. 2.24 indicates. It is important to notice that the width w does not enter either of the two models' resonant frequency, as it cancels out. The only two parameters that influence the resonant frequency of a constant rectangular cross section are the length l and thickness t.

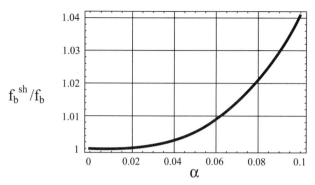

Figure 2.23 Long versus short microcantilevers: distribution function ratio in terms of thickness-to-length ratio (very close to the fixed root, $x = 0.99l$).

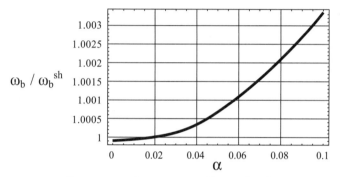

Figure 2.24 Long versus short microcantilevers: bending resonant frequency ratio in terms of thickness-to-length ratio.

2.2.4 Variable-cross-section members

The procedure of finding the resonant frequency via lumped-model stiffness and inertia fractions for microhinges and microcantilevers of constant cross section is now applied to members of variable cross sections. Configurations such as a trapezoid, or corner-filleted (with either circular or elliptic fillets) will be studied next. These configurations share the trait of being defined by a single geometric curve (line segment, circular or elliptic portion). It is assumed that the thin rectangular cross section is of variable width w and constant thickness t (except for one design where w is constant and t is variable), and that the material is homogeneous, which results in the material properties being constant. The representative lumped-parameter stiffness and inertia fractions, together with the corresponding resonant frequencies, are explicitly given here. They are going to be derived based on a generic model which is presented first, and which allows other geometric curves to be utilized for profile generation of different configurations. Similar approaches to this subject can be found in Lobontiu[2] and Lobontiu and Garcia.[3]

Generic formulation. As previously shown, the axial and torsional stiffnesses are found as the algebraic inverses of the corresponding compliances, and this relationship is also valid for the so-called definition bending stiffness (the one utilized in lumped-parameter resonant frequency calculations). It therefore suffices to determine compliances, which is a rather easier task for variable-cross-section members, compared to directly finding stiffnesses. As a consequence, the axial, torsional, and direct linear bending stiffnesses of a variable-cross-section microcantilever are calculated as

$$k_{a,e} = \frac{1}{C_{a,e}} = \frac{Et}{\int_0^l dx \,/\, [w(x)]}$$

$$k_{t,e} = \frac{1}{C_{t,e}} = \frac{3Gt^3}{\int_0^l dx \,/\, [w(x)]} \qquad (2.84)$$

$$k_{b,e} = \frac{1}{C_l} = \frac{12Et^3}{\int_0^l x^2 dx \,/\, [w(x)]}$$

Comparing the first two of Eqs. (2.84), we notice that

$$k_{t,e} = \frac{Gt^2}{3E} k_{a,e} \qquad (2.85)$$

and, consequently, the torsional stiffness can be found from the axial one.

It has also been shown that the shearing effects are important for short microcantilevers, and this translates to the direct linear bending compliance being calculated according to Eq. (2.28). As a direct result, the bending stiffness of a short microcantilever is

$$k_{b,e}^{sh} = \frac{1}{C_l^{sh}} = \frac{1}{C_l + \kappa(E\,/\,G)C_{a,e}} \qquad (2.86)$$

It can be shown, as also detailed in Lobontiu,[2] that the lumped-parameter mass which is dynamically equivalent to the distributed-parameter inertia of the microcantilever undergoing free axial vibrations can be calculated as

$$m_{a,e} = \rho t \int_0^l f_a^2(x) w(x) \, dx \qquad (2.87)$$

where the distribution function is calculated as

$$f_a(x) = \frac{C_a(x)}{C_{a,e}} \qquad (2.88)$$

with $\qquad C_a(x) = \dfrac{1}{Et} \displaystyle\int_x^l \dfrac{dx}{w(x)} \qquad (2.89)$

Similarly, the mechanical moment of inertia which is dynamically equivalent to the distributed inertia of a microcantilever during free torsional vibrations is

$$J_{t,e} = \frac{\rho t}{12} \int_0^l f_t^2(x) w(x) [w(x)^2 + t^2] \, dx \qquad (2.90)$$

where the torsion-related distribution function is calculated as

$$f_t(x) = \frac{C_t(x)}{C_{t,e}} \qquad (2.91)$$

with $\qquad C_t(x) = \dfrac{3}{Gt^3} \displaystyle\int_x^l \dfrac{dx}{w(x)} = \dfrac{3E}{Gt^2} C_a(x) \qquad (2.92)$

It can easily be checked that the distribution functions in axial and torsional vibrations of variable-cross-section microcantilevers are actually identical, and this conclusion extends to constant-cross-section microcantilevers, too. For a constant-cross-section microcantilever, the axial/torsional distribution function reduces to the one given in Eq. (2.48).

In bending, the effective mass is calculated as

$$m_{b,e} = \rho t \int_0^l f_b^2(x) w(x) \, dx \qquad (2.93)$$

The bending-related distribution function of Eq. (2.93) is calculated in Lobontiu[2] under the assumption that both a force and a moment that act at the microcantilever's free end produce bending. However, in defining the direct linear bending stiffness, only the effects of the end force are taken into account, and therefore it can be shown that the bending-related distribution function is calculated as

$$f_b(x) = \frac{C_l(x) - xC_c(x)}{C_l} \tag{2.94}$$

where the newly introduced compliances in the numerator are calculated as

$$C_l(x) = \frac{12}{Et^3}\int_x^l \frac{x^2\,dx}{w(x)} \qquad C_c(x) = \frac{12}{Et^3}\int_x^l \frac{x\,dx}{w(x)} \tag{2.95}$$

It can easily be demonstrated that for a constant rectangular cross-section microcantilever, the bending-related distribution function defined in Eqs. (2.94) and (2.95) reduces to Eq. (2.63).

In the case of short microcantilevers, where shearing effects and the corresponding deformations have to be accounted for, the bending-related distribution function is calculated as

$$f_b^{\mathrm{sh}}(x) = \frac{C_l^{\mathrm{sh}}(x) - xC_c(x)}{C_l^{\mathrm{sh}}} \tag{2.96}$$

where $\qquad C_l^{\mathrm{sh}}(x) = C_l(x) + \kappa\frac{E}{G}C_a(x) \tag{2.97}$

For a constant cross-section microcantilever, Eqs. (2.96) and (2.97) result in Eq. (2.77), which gives the bending-related distribution function in a separate (and independent) derivation.

Several variable-cross-section microcantilever configurations such as trapezoid or corner-filleted are analyzed next by providing lumped-parameter stiffness and inertia fractions together with the resonant frequencies corresponding to axial, torsional, and bending vibrations.

Trapezoid microcantilevers. A trapezoid configuration is shown in Fig. 2.25 together with the defining geometry. It is assumed that the microcantilever is fixed at its root and free at the opposite end, and that its constant thickness t is small (thin configuration). The variable width w, which is measured at a distance x from the free end, can be expressed as

$$w(x) = w_1 + \frac{(w_2 - w_1)x}{l} \tag{2.98}$$

As previously mentioned, the bending stiffness corresponding to point 1 in Fig. 2.25 and to deformation (rotation) about the y axis is calculated with the aid of the generic Eq. (2.84) and the width definition of Eq. (2.98). Thus, the bending stiffness is

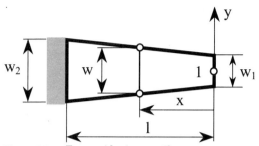

Figure. 2.25   Trapezoid microcantilever geometry.

$$k_{b,e} = \frac{Et^3(w_2 - w_1)^3}{6l^3[(w_1 - w_2)(3w_1 - w_2) + 2w_1^2 \ln(w_2/w_1)]} \tag{2.99}$$

When $w_2 \rightarrow w_1$, Eq. (2.98) reduces to Eq. (2.61), which gives the bending stiffness of a constant-cross-section microcantilever. The lumped mass corresponding to bending vibrations can be calculated by means of Eqs. (2.93), (2.94), and (2.98) and is found to be

$$m_{b,e} = \frac{215w_1 + 49w_2}{[560(w_1 + w_2]}m \tag{2.100}$$

Equation (2.100) reduces to Eq. (2.66)—which gives the bending mass of a constant-cross-section microcantilever—when $w_2 \rightarrow w_1$. The bending resonant frequency of the trapezoid microcantilever of Fig. 2.25 is, by way of the stiffness of Eq. (2.99) and the mass of Eq. (2.100),

$$\omega_{b,e} = 13.66 \sqrt{\frac{E(w_2 - w_1)}{\rho(215w_1 + 49w_2)[(w_1 - w_2)(3w_1 - w_2) + 2w_1^2\ln(w_2/w_1)]}}$$
$$\times \frac{t(w_2 - w_1)}{l^2} \tag{2.101}$$

The free axial vibrations are solved similarly by following the corresponding generic-approach algorithm. By using the first of Eqs. (2.84) coupled with Eqs. (2.87) and (2.98), the axial stiffness becomes

$$k_{a,e} = \frac{Et(w_2 - w_1)}{l \ln(w_2/w_1)} \tag{2.102}$$

For $w_2 \rightarrow w_1$, Eq. (2.102) transforms to Eq. (2.45), which yields the axial stiffness of a constant-cross-section microbar. The equivalent mass in axial vibration is determined by means of Eqs. (2.87), (2.88), and (2.98) as

$$m_{a,e} = \frac{3w_1 + w_2}{[6(w_1 + w_2)]} m \qquad (2.103)$$

Again, Eq. (2.103) reduces to Eq. (2.49) when $w_2 \to w_1$, as expected, because the trapezoid configuration transforms into a constant rectangular cross section to one for this limit condition. The axial vibration resonant frequency is therefore

$$\omega_{a,e} = \frac{2\sqrt{3}}{l} \sqrt{\frac{E(w_2 - w_1)}{\rho(3w_1 + w_2)\ln(w_2/w_1)}} \qquad (2.104)$$

The torsional resonant frequency is determined similarly with respect to the free end. The torsional stiffness is obtained from the axial one, according to Eq. (2.85). The inertia fraction corresponding to torsional vibrations is determined according to Eqs. (2.90), (2.91), (2.92), and (2.98) as

$$J_{t,e} = \frac{\rho l t [10w_1^3 + 6w_1^2 w_2 + 3w_1 w_2^2 + w_2^3 + 5t^2(3w_1 + w_2)]}{720} \qquad (2.105)$$

When $w_2 \to w_1$, Eq. (2.105) changes to Eq. (2.55), which defines the torsional moment of inertia for a constant-cross-section microbar. The torsional resonant frequency can be calculated as

$$\omega_{t,e} = \frac{4\sqrt{15}t}{l} \sqrt{\frac{G(w_2 - w_1)}{\rho[10w_1^3 + 6w_1^2 w_2 + 3w_1 w_2^2 + w_2^3 + 5t^2(3w_1 + w_2)]\ln(w_2/w_1)}} \qquad (2.106)$$

Another trapezoid microcantilever is sketched in Fig. 2.26a and b where the width w is constant and the thickness varies linearly from t_1 at the free end to t_2 at the fixed root.

The axial stiffness of this microcantilever is

$$k_{a,e} = \frac{Ew(t_2 - t_1)}{l \ln(t_2/t_1)} \qquad (2.107)$$

When $t_2 \to t_1$, Eq. (2.107) reduces to Eq. (2.45), which gives the axial stiffness of a constant rectangular cross-section microcantilever.

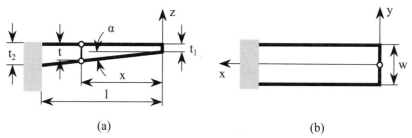

(a)   (b)

Figure 2.26 Variable-thickness trapezoid microcantilever: (a) side view; (b) top view.

The lumped-parameter mass which is dynamically equivalent to the distributed-parameter inertia of this microcantilever is calculated by considering two possibilities of defining the distribution function, namely, first, as provided by the generic Eq. (2.88), which takes into consideration that the cross section is variable and, second, by using Eq. (2.48), which assumes the cross section is constant. In the first variant, the two compliances defining the distribution function of Eq. (2.88) are

$$C_a(x) = \frac{1}{Ew}\int_x^l \frac{dx}{t(x)} \qquad C_{a,e} = \frac{1}{Ew}\int_0^l \frac{dx}{t(x)} \tag{2.108}$$

The lumped-parameter mass is calculated as

$$m_{a,e} = \rho w \int_0^l t(x) f_a^2(x)\,dx \tag{2.109}$$

In both Eqs. (2.108) and (2.109), the variable thickness is determined as

$$t(x) = t_1 + \frac{t_2 - t_1}{l}x \tag{2.110}$$

The effective mass is

$$m_{a,e} = \frac{\rho l w\{t_2^2 - t_1^2 - 2t_1^2[1 + \ln(t_2/t_1)]\ln(t_2/t_1)\}}{4(t_2 - t_1)\ln^2(t_2/t_1)} \tag{2.111}$$

This equation reduces to Eq. (2.49), which provides the effective mass of a constant rectangular cross-section microcantilever, when $t_2 \to t_1$.

As mentioned previously, the effective mass in axial vibrations can also be calculated by using the distribution function of Eq. (2.48)—which corresponds to a constant-cross-section member—instead of the distribution function that has just been used. The objective here is to

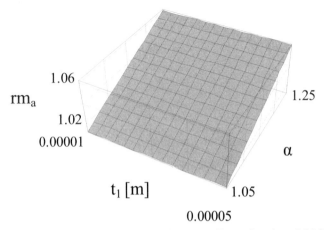

rm$_a$ 1.06   1.25

1.02

0.00001

α

t$_1$ [m]   1.05

0.00005

Figure 2.27   Effective mass ratio [Eq. (2.113)] as a function of thickness parameters.

assess the errors between the two models that are involved in mass calculations. In this case, the effective mass has the simpler expression:

$$m_{a,e}^{*} = \frac{\rho l\, w(3t_1 + t_2)}{12} \qquad (2.112)$$

This equation, too, reduces to Eq. (2.49) when $t_2 \to t_1$, which proves its validity. If the mass ratio is analyzed

$$rm_a = \frac{m_{a,e}^{*}}{m_{a,e}} \qquad (2.113)$$

by considering that

$$t_2 = \alpha t_1 \qquad (2.114)$$

the plot of Fig. 2.27 can be drawn.

As Fig. 2.27 indicates, the errors that are generated when the effective mass is calculated by using the distribution function corresponding to a constant-cross-section member, instead of the proper distribution function defining a variable-cross-section one, are quite small; and therefore using the simpler Eq. (2.112) instead of the exact and more complex Eq. (2.111) is sufficiently accurate. Moreover, when the resonant frequency needs to be calculated, the errors that are set up by using the simplified effective mass are further reduced as the resonant frequency depends on the square root of the effective mass. For instance, an error of 6 percent in the effective mass will translate to a 2.45 percent error in the resonant frequency, which is quite acceptable. As a consequence, one gets the simplified resonant frequency corresponding to axial vibrations by combining Eqs. (2.107) and (2.112):

$$\omega_{a,e}^* = \frac{3.46}{l}\sqrt{\frac{E(t_2 - t_1)}{\rho(3t_1 + t_2)\ln(t_2/t_1)}} \qquad (2.115)$$

In torsion, the stiffness corresponding to the free end of the trapezoid microcantilever of Fig. 2.26 is found by means of the axial stiffness [Eq. (2.107)] and the connection Eq. (2.85).

Again, the lumped inertia will be calculated by using the two variants utilized in determining the effective axial inertia, as performed above. The effective moment of inertia that results by using the exact distribution function of Eq. (2.91) is

$$J_{t,e} = \frac{\rho l w\left\{(t_2^2 - t_1^2)(t_1^2 + t_2^2 + 8w^2) - 4t_1^2\ln(t_2/t_1)[t_1^2 + 4w^2 + 2(t_1^2 + 2w^2)\ln(t_2/t_1)]\right\}}{384(t_2 - t_1)\ln^2(t_2/t_1)} \qquad (2.116)$$

Similarly, the effective mechanical moment of inertia calculated with the constant-cross-section distribution function of Eq. (2.48) is

$$J_{t,e}^* = \frac{\rho l w[10t_1^3 + 6t_1^2 t_2 + 3t_1 t_2^2 + t_2^3 + 5(3t_1 + t_2)w^2]}{720} \qquad (2.117)$$

The errors that are generated by using the approximate Eq. (2.117) instead of the exact Eq. (2.116) can be assessed by following the path utilized in the previous comparison corresponding to the effective mass in axial vibrations, and the conclusions are very similar. Note that, again, both Eqs. (2.116) and (2.117) reduce to Eq. (2.55) when $t_2 \to t_1$, so they are both valid. As a consequence, the torsion-related resonant frequency of the microcantilever of Fig. 2.26 can be calculated as

$$\omega_{t,e}^* = \frac{21.91t_1 t_2\sqrt{G/\left\{\rho(t_1 + t_2)[10t_1^3 + 6t_1^2 t_2 + 3t_1 t_2^2 + t_2^3 + 5(3t_1 + t_2)w^2]\right\}}}{l} \qquad (2.118)$$

The lumped-parameter stiffness corresponding to the out-of-plane bending of the trapezoid microcantilever of Fig. 2.26 is

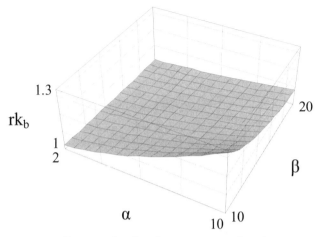

$\mathrm{rk_b}$

Figure 2.28   Resonant bending frequency comparison: long- versus short-beam models.

$$k_{b,e} = \frac{E w t_2^2 (t_2 - t_1)^3}{6 l^3 [t_2^2 (2 \ln(t_2/t_1) - 3) + 4 t_1 t_2 - t_1^2]} \tag{2.119}$$

When $t_2 \to t_1$, Eq. (2.119), simplifies to Eq. (2.61), which provides the bending stiffness of a constant rectangular cross-section microcantilever; this proves the validity of Eq. (2.119).

Example: Evaluate the bending stiffness of the trapezoid microcantilever of Fig. 2.26 by comparing the results of the long-beam model to the ones generated by the short-beam model.

The short-beam model bending stiffness is calculated by means of Eqs. (2.21), (2.28), and (2.110). By considering the substitutions

$$t_2 = \alpha t_1 \quad l = \beta t_1 \tag{2.120}$$

the following bending stiffness ratio is formulated just in terms of the two nondimensional parameters of Eq. (2.120):

$$rk_b = \frac{k_{b,e}}{k_{b,e}^{sh}} \tag{2.121}$$

which is plotted in Fig. 2.28.

As this figure indicates, the stiffness obtained from the long-beam model can be 1.3 times larger than the similar stiffness calculated by means of the short-beam model, particularly for relatively short and thick configurations.

To calculate the exact effective inertia corresponding to free bending vibrations, the distribution function of Eq. (2.94) has to be utilized, where

$$C_l(x) = \frac{12}{Ew} \int_x^l \frac{x^2 dx}{t(x)^3}$$

$$C_l = \frac{12}{Ew} \int_0^l \frac{x^2 dx}{t(x)^3} \tag{2.122}$$

$$C_c(x) = \frac{12}{Ew} \int_x^l \frac{x\, dx}{t(x)^3}$$

The effective mass is calculated as

$$m_{b,e} = \rho w \int_0^l t(x) f_b^2(x)\, dx \tag{2.123}$$

After we perform the calculations necessitated in Eq. (2.123), the bending-related effective mass is

$$m_{b,e} = \frac{\rho l w \left\{ (t_2 - t_1)(9t_1^5 - 75t_1^4 t_2 + 197t_1^3 t_2^2 - 155t_1^2 t_2^3 \\ -44t_1 t_2^4 + 8t_2^5) - 12t_1^2 t_2^2 \left\{ 6t_2^2(\ln^2 t_1 - \ln^2 t_2) + [\, 12t_2^2 \ln t_2 \\ -(2t_1 - 7t_2)(2t_1 - 3t_2)]\ln(t_1/t_2) \right\} \right\}}{36(t_2 - t_1)[(3t_2 - t_1)(t_2 - t_1) + 2t_2^2 \ln(t_2/t_1)]^2} \tag{2.124}$$

which is quite involved. The effective bending mass can also be calculated by using the distribution function of Eq (2.63), which corresponds to a constant-cross-section microcantilever, instead of the exact distribution function of Eq. (2.94). In doing so, the following effective mass is obtained:

$$m_{b,e}^* = \frac{\rho l w (215t_1 + 49t_2)}{1120} \tag{2.125}$$

which is considerably simpler than what Eq. (2.124) yielded. Both Eqs. (2.124) and (2.125) reduce to Eq. (2.66) which gives the effective bending inertia of a constant-cross-section microcantilever when $t_2 \to t_1$. Again, a comparison has been performed between the effective masses provided by the two distribution functions through analyzing the mass ratio:

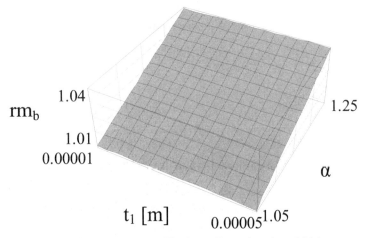

$$\text{rm}_b \quad \begin{matrix} 1.04 \\ 1.01 \\ 0.00001 \end{matrix} \qquad\qquad\qquad \begin{matrix} 1.25 \\ \\ \alpha \end{matrix}$$

$$t_1 \, [m] \qquad 0.00005^{1.05}$$

Figure 2.29   Effective mass ratio [Eq. (2.126)] as a function of thickness parameters.

$$rm_b = \frac{m_{b,e}^*}{m_{b,e}} \qquad\qquad (2.126)$$

Figure 2.29 plots the ratio of Eq. (2.126) in terms of t_1 and α, as defined in Eq. (2.114). As this figure suggests, the errors induced by using the approximation mentioned above in calculating the bending-related effective inertia are quite small, and they decrease further when one is using the same approximation in determining the bending resonant frequency, as the frequency depends on the square root of the effective mass. As a consequence, the approximate bending resonant frequency is given by

$$\omega_{b,e}^* = \frac{13.66 t_2 (t_2 - t_1)}{l^2} \sqrt{\frac{E(t_2 - t_1)}{\rho \, (215 t_1 + 49 t_2)[(2\ln(t_2/t_1) - 3)t_2^2 + 4 t_1 t_2 - t_1^2]}} \qquad (2.127)$$

Following the conclusions derived after analyzing the errors between the resonant frequency results produced by the two types of distribution functions, the ones corresponding to constant-cross-section members— Eqs. (2.48) and (2.63)—will be utilized from this point on.

Corner-filleted microcantilevers.   Corner-filleted flexible components are utilized to both mitigate the effects of stress concentration at sharp corners, especially in the case of microresonators, where repetitive operation under resonant conditions leads to fatigue, which is the root

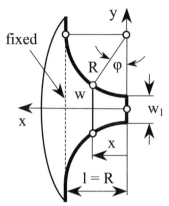

Figure 2.30 Circularly filleted microcantilever.

candidate cause for mechanical failure, and to offer a solution of augmenting the design space through variation of some key geometric parameters, such as the shape and dimensions of the fillet areas. Two filleted configurations are studied next, namely the circularly and the elliptically filleted microcantilevers. Both can be used as either standalone components or basic units in more complex shapes of flexible connectors.

Circularly filleted microcantilever. The top view of a thin, constant-thickness, circularly filleted microcantilever (whose compliances are derived in Lobontiu and Garcia[3]) is sketched in Fig. 2.30. Its rather sturdy configuration is particularly suited for force applications such as atomic force microscopy (AFM). The width w varies along the length, and its value can be expressed as

$$w = w_1 + 2R(1 - \cos \varphi) \qquad (2.128)$$

The abscissa x can also be expressed in terms of the variable angle φ (the limits to φ are $0°$ and $90°$ when x ranges from 0 to R) as

$$x = R \sin \varphi \qquad (2.129)$$

The axial stiffness of the fixed-free circularly filleted microcantilever can be expressed as

$$k_{a,e} = \frac{Et}{\sqrt{\dfrac{(1 + 2R/w_1)(2R + w_1)}{\big/(4R + w_1)\arctan\sqrt{1 + 4R/w_1} - \pi/4}}} \qquad (2.130)$$

The lumped mass which needs to be placed at the free end of the microcantilever and which is equivalent to the distributed inertia of the component during the free axial vibrations is

$$m_{a,e} = \rho R t \left(0.036R + \frac{w_1}{3} \right) \tag{2.131}$$

The resonant frequency corresponding to free axial vibrations is

$$\omega_{a,e} = 2 \sqrt{ \begin{array}{l} E / \{ \rho R (0.036R + w_1/3) \\ \times [\sqrt{(1 + 2R/w_1)(2R + w_1)} / (4R + w_1) \\ \times \arctan\sqrt{1 + 4R/w_1} - \pi/4] \} \end{array} } \tag{2.132}$$

The torsional stiffness is expressed as a function of the axial one according to Eq. (2.85). The equivalent mechanical moment of inertia of the mass that is located at the microcantilever's free end is

$$J_{t,e} = \frac{\rho R t [0.007R^3 + 0.038R^2 w_1 + 0.333 w_1 (w_1^2 + t^2) + 0.036R(3w_1^2 + t^2)]}{12} \tag{2.133}$$

The torsion-related resonant frequency is

$$\omega_{t,e} = \frac{2t\sqrt{\begin{array}{l} G / \{ \rho R [0.007R^3 + 0.038R^2 w_1 \\ + 0.333 w_1 (w_1^2 + t^2) + 0.036R(3w_1^2 + t^2)] \} \end{array}}}{\sqrt{\begin{array}{l} (1 + 2R/w_1)(2R + w_1) \\ / (4R + w_1) \arctan\sqrt{1 + 4R/w_1} - \pi/4 \end{array}}} \tag{2.134}$$

In bending, two variants are treated corresponding to the long- and short-beam models. For relatively long configurations, the lumped stiffness and equivalent mass are calculated according to the Euler-Bernoulli model, whereas for short configurations the lumped parameters are determined by means of Timoshenko's model and consideration of the shearing effects.

For long microcantilevers, the bending-related stiffness is

$$k_{b,e} = \frac{4Et^3}{\{3[14.283R^2 + 16.566Rw_1 + 3.141w_1^2 \\ -4(2R + w_1)\sqrt{w_1(4R + w_1)}\arctan\sqrt{1 + 4R/w_1}]\}} \tag{2.135}$$

The lumped mass which is equivalent to the distributed inertia of the microcantilever undergoing free bending vibrations is

$$m_{b,e} = \rho Rt(0.014R + 0.236w_1) \tag{2.136}$$

The bending-related resonant frequency of the long circularly filleted microcantilever is

$$\omega_{b,e} = \frac{1.15t\sqrt{E/[\rho R(0.014R + 0.236w_1)]}}{\sqrt{\begin{array}{c}14.283R^2 + 16.566Rw_1 + 3.141w_1^2 \\ -4(2R + w_1)\sqrt{w_1(4R + w_1)}\arctan\sqrt{1 + 4R/w_1}\end{array}}} \tag{2.137}$$

For short microcantilevers, the Timoshenko model with consideration of the shearing effects has to be utilized, as detailed in previous sections. It has been shown that the linear bending compliance C_l as well as the axial compliance C_a is needed to enable determination of the linear shear-dependent compliance according to Eq. (2.28). These two compliances can be calculated by means of their definition Eqs. (2.25) and (2.27); their explicit forms, as mentioned previously, are given in Lobontiu and Garcia[3] and are not reproduced here.

It can be shown that the lumped-parameter stiffness for the Timoshenko model is

$$k_{b,e}^{sh} = \frac{4EGt^3}{\begin{array}{c}-3.14\kappa Et^2 + G(42.85R^2 + 49.7Rw_1 + 9.42w_1^2) \\ +[4\kappa Et^2(2R + w_1) - 12(8R^2 + 6Rw_1 \\ +w_1^2)w_1G]\arctan\sqrt{1 + 4R/w_1}/\sqrt{w_1(4R + w_1)}\end{array}} \tag{2.138}$$

and that the lumped mass is

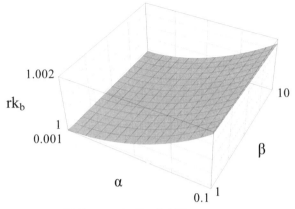

rk_b

Figure 2.31   Stiffness ratio, Eq. (2.141).

$$\rho R t [140\kappa^2 E^2 t^4 (0.88R + 8w_1)$$
$$+56\kappa E G R^2 t^2 (10.4R + 132w_1)$$
$$m_{b,e}^{sh} = \frac{+ G^2 R^4 (765.4R + 12{,}672w_1)]}{3360(4GR^2 + \kappa E t^2)^2} \tag{2.139}$$

The corresponding resonant frequency, which is too complex to be given here, can be determined by combining Eqs. (2.138) and (2.139), according to its definition.

Example: Analyze the effects of shearing on the bending resonant frequency of a circularly filleted microcantilever.

The following nondimensional parameters are introduced:

$$\alpha = \frac{t}{R} \quad \beta = \frac{w_1}{R} \tag{2.140}$$

With their help, the following ratios are constructed:

$$rk_b = \frac{k_{b,e}}{k_{b,e}^{sh}} \tag{2.141}$$

$$rm_b = \frac{m_{b,e}^{sh}}{m_{b,e}} \tag{2.142}$$

These depend on only the nondimensional parameters α and β [a value of $\kappa = \frac{5}{6}$ has been chosen and Eq. (2.70) has been used for the connection between E and G, with Poisson's ratio $\mu = 0.25$]. Figures 2.31 and 2.32 are

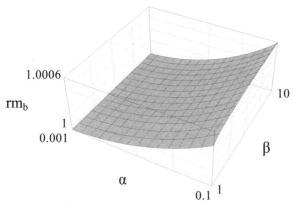

Figure 2.32 Effective mass ratio, Eq. (2.142).

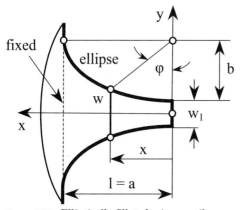

Figure 2.33 Elliptically filleted microcantilever.

three-dimensional plots of the ratios defined in Eqs. (2.141) and (2.142). As both Figs. 2.31 and 2.32 indicate, the shearing effects on both the stiffness and the effective mass are reduced, almost negligible over the parameter ranges. It is interesting to remark that while the stiffness yielded by the short-beam model is smaller than the one produced by the long-beam model, this relationship reverses in terms of effective masses. It is therefore evident that the bending resonant frequencies of the two models are very close to each other.

Elliptically filleted microcantilever. Another filleted microcantilever config-uration, also mentioned by Lobontiu and Garcia[3] in terms of its com-pliances, is the elliptical one whose top view with the defining geometry is sketched in Fig. 2.33. This design is formed of two identical elliptic portions, which are defined by the two semiaxes a and b. The longer semiaxis in Fig. 2.33 is equal to the length l of the microcantilever.

The variable width w is expressed in terms of the minimum width w_1, the shorter semiaxis length b, and the variable angle φ as

$$w(x) = w_1 + 2b(1 - \cos \varphi)\qquad(2.143)$$

The abscissa x is expressed as

$$x = a \sin \varphi\qquad(2.144)$$

and it spans the $[0, a]$ interval when φ ranges between $0°$ and $90°$. The axial stiffness is

$$k_{a,e} = \frac{4Ebt}{a[4(2b + w_1)/\sqrt{w_1(4b + w_1)}\arctan\sqrt{1 + 4b/w_1} - \pi]}\qquad(2.145)$$

When $a \to R$ and $b \to R$, Eq. (2.145) changes to Eq. (2.130), which expresses the lumped-parameter axial stiffness of a circularly filleted microcantilever.

The lumped-parameter effective mass corresponding to free axial vibrations is

$$m_{a,e} = \rho at\left(0.036b + \frac{w_1}{3}\right)\qquad(2.146)$$

When $a \to R$ and $b \to R$, Eq. (2.146) transforms to Eq. (2.131), which gives the effective mass of a circularly filleted microcantilever. The axially related resonant frequency is therefore

$$\omega_{a,e} = \frac{2}{a}\sqrt{\frac{Eb}{\rho(0.036b + w_1/3)[4(2b + w_1)}{/\sqrt{w_1(4b + w_1)}\arctan\sqrt{1 + 4b/w_1} - \pi]}}\qquad(2.147)$$

The torsional stiffness is obtained from the axial one according to Eq. (2.85). The effective mechanical moment of inertia corresponding to free torsional vibrations is

$$J_{t,e} = \frac{\rho at[0.007b^3 + 0.038b^2 w_1 + 0.333w_1(t^2 + w_1^2)}{+0.036b(t^2 + 3w_1^2)]}{12}\qquad(2.148)$$

For $a \to R$ and $b \to R$, Eq. (2.148) changes to Eq. (2.133), which gives the effective mechanical moment of inertia of a circularly filleted microcantilever. The torsional resonant frequency is

$$
\omega_{t,e} = \frac{4t}{a} \frac{\sqrt{\dfrac{bG}{\rho\,[\,0.007b^3 + 0.038b^2 w_1 + 0.333 w_1 (t^2 + w_1^2) + 0.036b(t^2 + 3w_1^2)\,]}}}{\sqrt{4(2b + w_1)/\sqrt{w_1(4b + w_1)}\,\arctan\sqrt{1 + 4b/w_1} - \pi}} \tag{2.149}
$$

The lumped-parameter out-of-the-plane bending stiffness of a relatively long, elliptically filleted microcantilever is

$$
k_{b,e} = \frac{4Eb^3 t^3}{3a^3[\,14.283b^2 + 16.566bw_1 + 3.14w_1^2 \\ -4(2b + w_1)\sqrt{w_1(4b + w_1)}\,\arctan\sqrt{1 + 4b/w_1}\,]} \tag{2.150}
$$

Equation (2.150) reduces to Eq. (2.135), which defines a circularly filleted microcantilever, when $a \to R$ and $b \to R$.

For short configurations, the bending stiffness of this design is

$$
k_{b,e}^{sh} = \frac{4EGb^3 t^3}{a\{-\pi\kappa Eb^2 t^2 + 3Ga^2[2(4 + \pi)b^2 + 4(1 + \pi)bw_1 \\ + \pi w_1^2] - 4(2b + w_1)[-\kappa Eb^2 t^2 \\ + 3Ga^2 w_1(4b + w_1)\arctan\sqrt{1 + 4b/w_1}\,]/\sqrt{w_1(4b + w_1)}\}} \tag{2.151}
$$

and it can be seen that for $a \to R$ and $b \to R$, this equation becomes Eq. (2.138), which characterizes a circularly filleted microcantilever.

The effective bending mass for a long microcantilever is

$$
m_{b,e} = \rho at(0.014b + 0.236w_1) \tag{2.152}
$$

which transforms to Eq. (2.136) when $a \to R$ and $b \to R$.

The resonant frequency of a long, elliptically filleted design is determined by means of Eqs. (2.150) and (2.152) as

$$
\omega_{b,e} = \frac{2bt}{\sqrt{3}a^2} \frac{\sqrt{\dfrac{bE}{\rho(0.014b + 0.236w_1)}}}{\sqrt{2(4 + \pi)b^2 + 4(1 + \pi)bw_1 - 4(2b + w_1)\sqrt{w_1(4b + w_1)}\,\arctan\sqrt{1 + 4b/w_1}}} \tag{2.153}
$$

The effective bending mass for a short microconfiguration is

$$\rho at[140\kappa^2 E^2 t^4(0.876b + 8w_1)$$

$$+56\kappa EGa^2 t^2(10.4b + 132w_1)$$

$$m_{b,e}^{sh} = \frac{+G^2 a^4(765.4b + 12{,}672w_1)]}{3360(46a^2 + \kappa Et^2)^2}$$

(2.154)

and this equation changes to the one defining a circularly filleted microcantilever [Eq. (2.139)] when $a \to R$ and $b \to R$.

The bending resonant frequency of a relatively short configuration can be determined by means of Eqs. (2.151) and (2.154) and is not explicitly given here.

Example: Compare the regular bending resonant frequency to the shearing-dependent one for an elliptic filleted microcantilever of rectangular cross section with $b = 2a$.

The following substitutions are introduced:

$$a = \alpha t \qquad w_1 = \beta t$$

(2.155)

such that the frequency ratio

$$r\omega = \frac{\omega_{b,e}}{\omega_{b,e}^{sh}}$$

(2.156)

becomes only a function of α and β. Figure 2.34 is the three-dimensional plot of $r\omega$.

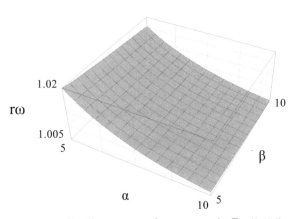

Figure 2.34 Bending resonant frequency ratio, Eq. (2.156).

As the situation was with the circular filleted microcantilever, for this configuration, too, the influence of shearing on the bending resonant frequency is reduced, as the ratio of Eq. (2.156) is slightly larger than 1, even for very short and wide configurations (small values of α and β).

As shown in Chap. 3, microcantilevers and/or microhinges of more complex configurations can be formed by serially combining the simple shapes that have been presented thus far, such as constant-cross-section members, and circularly or elliptically filleted designs. Finding the representative lumped-parameter properties of those compound members will imply utilization of the elastic and inertia properties of the basic designs presented thus far in this chapter.

2.3 Distributed-Parameter Modeling and Design

The distributed-parameter modeling approach describes the vibratory motion of a mechanical system by formulating partial differential equations (PDEs) that reflect the system's behavior in both time and space. As a consequence, it becomes possible to directly evaluate the resonant characteristics of a system without having to separately calculate stiffness and inertia fractions to obtain a specified resonant frequency—as was the case with the lumped-parameter modeling approach.

Microcantilevers and microbridges of constant cross section (generally rectangular) are analyzed first, and then the first resonant frequencies are calculated. As shown when we examined the same aspect by the lumped-parameter procedure, the first resonant frequency corresponds to bending about the sensitive axis. The solution to the partial differential equation (PDE) can be expressed uniquely for both microcantilevers and microbridges. The different boundary conditions will discriminate between the solutions particular to each of the two components. The relevant resonant frequencies will also be derived for circular rings, thin plates, and membranes by both exact integration methods and approximate ones.

2.3.1 Line micromembers

Long configurations (treated by means of the Euler-Bernoulli model) and short ones (modeled by means of Timoshenko's assumption) are analyzed in this section, and the bending resonant frequencies corresponding to cantilevers and bridges are derived for constant-cross-section members.

Long microcantilevers and bridges.   The differential equation that governs the free vibrations of a long variable-cross-section beam (Euler-Bernoulli model) is

$$\frac{\partial^4[EI_y(x)u_z(x, t)]}{\partial x^4} + \frac{\partial^2[\rho A(x)u_z(x, t)]}{\partial t^2} = 0 \qquad (2.157)$$

When the cross section is constant, Eq. (2.157) simplifies to

$$EI_y \frac{\partial^4[u_z(x, t)]}{\partial x^4} + \rho A \frac{\partial^2[u_z(x, t)]}{\partial t^2} = 0 \qquad (2.158)$$

The method of separation of variables is usually employed in solving the problem of free vibrations which assumes that

$$u_z(x, t) = U_z(x)T(t) \qquad (2.159)$$

By using the notation $u_z(x)$ for $U_z(x)$, the following differential equation can be derived from Eqs. (2.158) and (2.159):

$$\frac{d^4u_z(x)}{dx^4} - \beta^4 u_z(x) = 0 \qquad (2.160)$$

where
$$\beta^4 = \frac{\rho A \omega^2}{EI_y} \qquad (2.161)$$

The general solution to Eq. (2.161) is, as shown by Thomson,[6] for instance, of the form:

$$u_z(x) = A \cosh(\beta x) + B \sinh(\beta x) + C \cos(\beta x) + D \sin(\beta x) \qquad (2.162)$$

For a cantilever of length l and with fixed-free ends, the following equation is obtained by imposing the zero slope and zero deflection conditions at the fixed end:

$$1 + \cosh(\beta l) \cos(\beta l) = 0 \qquad (2.163)$$

By solving the transcendental Eq. (2.163), the following solutions are obtained, which describe the first three modes:

$$\beta l = \begin{cases} 1.8751 \\ 4.6941 \\ 7.85476 \end{cases} \qquad (2.164)$$

By combining the definition of β [Eq. (2.161)] and the value of βl corresponding to the first mode [Eq. (2.164)], the natural (first resonant) frequency is

$$\omega = 3.52 \sqrt{\frac{EI_y}{ml^3}} \tag{2.165}$$

A similar reasoning is applied to a microbridge (fixed-fixed member), and by enforcing the zero slope and zero deflection conditions at the two ends, the following equation is obtained:

$$1 - \cosh(\beta l)\cos(\beta l) = 0 \tag{2.166}$$

The numerical values of βl corresponding to the first three modes are

$$\beta l = \begin{cases} 4.73004 \\ 7.8532 \\ 10.9956 \end{cases} \tag{2.167}$$

Again, the natural frequency corresponding to the first mode is

$$\omega = 22.373 \sqrt{\frac{EI_y}{ml^3}} \tag{2.168}$$

Short microcantilevers and bridges.   The differential Eqs. (2.72) describing the behavior of short beams by means of Timoshenko's theory can be written into the alternative form:

$$EI_y \frac{d^3\theta_y(x)}{dx^3} - \rho A\omega^2 u_z(x) = 0$$

$$\theta_y(x) - \frac{du_z(x)}{dx} = \frac{EI_y}{\kappa AG}\frac{d^2\theta_y(x)}{dx^2} \tag{2.169}$$

By combining Eqs. (2.169) the following differential equation is obtained:

$$\frac{d^4\theta_y(x)}{dx^4} + \gamma^2 \frac{d^2\theta_y(x)}{dx^2} - \beta^4\theta_y(x) = 0 \tag{2.170}$$

with β given in Eq. (2.61) and γ defined as

$$\gamma^2 = \frac{\rho\omega^2}{\kappa G} \tag{2.171}$$

The solution to Eq. (2.170) is of the form:

$$\theta_y(x) = A\cosh(r_1 x) + B\sinh(r_1 x) + C\cos(r_2 x) + D\sin(r_2 x) \qquad (2.172)$$

where $\qquad r_1^2 = \dfrac{\sqrt{\gamma^4 + 4\beta^4} - \gamma^2}{2} \qquad r_2^2 = \dfrac{\sqrt{\gamma^4 + 4\beta^4} + \gamma^2}{2} \qquad (2.173)$

The deflection is calculated from the first of Eqs. (2.169) as

$$u_z(x) = \frac{\begin{aligned}&r_1^3[A\sinh(r_1 x) + B\cosh(r_1 x)]\\ &+ r_2^3[C\sin(r_2 x) - D\cos(r_2 x)]\end{aligned}}{\beta^4} \qquad (2.174)$$

The natural frequencies of relatively short line members can be determined by using specific boundary conditions for fixed-free beams (microcantilevers) and fixed-fixed beams (microbridges), as indicated for long line members. Numerical examples are not included here contrasting the results provided by the long- and short-beam model predictions, but it can easily be verified that small differences exist between the two models' bending-related resonant frequencies. This topic was discussed earlier in this chapter by using lumped-parameter models.

Long members with axial load.   In the case where an axial load acts at the free end of a microcantilever, for instance, as shown in Fig. 2.35, the bending natural frequency will change from its regular value due to a change in the elastic potential energy of the member.

The elemental deformation ds is related to its projections dx and du_z as

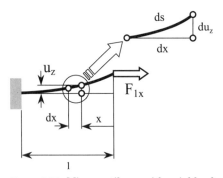

Figure 2.35   Microcantilever with axial load.

$$ds^2 = dx^2 + du_z^2 \tag{2.175}$$

The increase in length due to the action of the axial force can be expressed as

$$ds - dx = \frac{du_z^2}{ds + dx} \cong \frac{1}{2}\left(\frac{du_z}{dx}\right)^2 dx \tag{2.176}$$

The work performed by the axial force is converted to strain energy and is equal to

$$U_a = \frac{F_x}{2}\int_0^l \left(\frac{du_z}{dx}\right)^2 dx \tag{2.177}$$

and therefore the total strain energy (the sum of bending and axial strain energy contributions) is

$$U = U_b + U_a = \frac{EI_y}{2}\int_0^l \left(\frac{d^2 u_z}{dx^2}\right)^2 dx + \frac{F_x}{2}\int_0^l \left(\frac{du_z}{dx}\right)^2 dx \tag{2.178}$$

Assuming the deflection varies according to the sinusoidal law u_z sin (ωt), the kinetic energy of the vibrating microcantilever is

$$T = \frac{\omega^2 \rho A}{2}\int_0^l u_z^2 dx \tag{2.179}$$

Two approximate methods are presented next which are designed to determine the bending-related resonant frequency of a microcantilever under the action of a tip axial load.

Rayleigh's procedure. According to Rayleigh (see Timoshenko,[5] for instance) the natural frequency can be found by enforcing $U = T$ and by considering a certain shape of the deformed cantilever. When the deflection curve is the one produced by a force F_{1z} acting at the free tip, namely,

$$u_z(x) = \frac{F_{1z} l^3}{3EI_y}\left[1 - \frac{3x}{2l} + \frac{x^3}{(2l)^3}\right] \tag{2.180}$$

the altered resonant frequency becomes

$$\omega = 3.56753\sqrt{\left(1 + \frac{2F_x l^2}{5EI_y}\right)\frac{EI_y}{ml^3}} \tag{2.181}$$

Equation (2.181) reduces to an expression similar to that provided by Eq. (2.165) when $F_x = 0$.

A better approximation of the natural frequency can be obtained when we consider that the deflection curve is generated by a uniformly distributed load q_z acting over the entire length of the cantilever, which is given as

$$u_z = \frac{q_z(x^4 - 4l^3x + 3l^4)}{24EI_y} \qquad (2.182)$$

In this case the natural frequency is

$$\omega = 3.53\sqrt{\left(1 + \frac{5F_xl^2}{14EI_y}\right)\frac{EI_y}{ml^3}} \qquad (2.183)$$

Rayleigh-Ritz procedure. Rayleigh's procedure enabled us to find the first resonant frequency (in bending, in this case, but torsion and axial loading can also be dealt with). Ritz proposed an algorithm, based on Rayleigh's method, that enabled finding of the higher resonant frequencies, in addition to the first one. According to the Rayleigh-Ritz approximate procedure, the natural frequency can be found by enforcing $U = T$ and by choosing a deflection of the form:

$$u_y = a_0\varphi_0(x) + a_1\varphi_1(x) + \cdots + a_n\varphi_n(x) + \cdots \qquad (2.184)$$

where $\varphi_0(x)$, $\varphi_1(x)$, ..., $\varphi_n(x)$ are shape functions that comply with the (fixed-free) boundary conditions of the microcantilever, and the coefficients a_0, a_1, ..., a_n are the unknowns of the problem. In truncating the potentially infinite series of Eq. (2.184), one imposes limitations on the deflection curve; as a consequence, the resonant frequencies will always be higher than the real ones. To get smaller resonant frequencies, the Rayleigh-Ritz procedure selects the coefficients a_0, a_1, ..., a_n so as to make the function

$$\omega^2 = \frac{EI_y\int_0^l (d^2u_z/dx^2)^2dx + F_{1x}\int_0^l (du_z/dx)^2dx}{\rho A\int_0^l u_z^2dx} \qquad (2.185)$$

a minimum. This condition requires

$$\frac{\partial\omega^2}{\partial a_i} = 0 \quad i = 1, 2, \ldots \qquad (2.186)$$

The generic Eq. (2.186) can be written in the form:

$$\sum_i c_{ij} a_i = 0 \quad j = 1, 2, \ldots \tag{2.187}$$

where $c_{ij} = c_{ji}$. This represents a set of homogeneous equations in a_i whose solution is nontrivial when the system's determinant is zero, namely,

$$\det \begin{bmatrix} c_{11} & c_{12} & \cdots & c_{1n} \\ c_{12} & c_{22} & \cdots & c_{2n} \\ \cdots & \cdots & \cdots & \cdots \\ c_{1n} & c_{2n} & \cdots & c_{nn} \end{bmatrix} = 0 \tag{2.188}$$

Equation (2.188), the characteristic equation, is an algebraic equation of the nth degree in ω^2 which will provide the first n resonant frequencies (all bending-related) of the microcantilever.

The two shape functions

$$\varphi_1(x) = \left(1 - \frac{x}{l}\right)^2 \quad \varphi_2(x) = \frac{x}{l}\left(1 - \frac{x}{l}\right)^2 \tag{2.189}$$

satisfy the boundary conditions of a fixed-free beam, which are zero shearing force and zero bending moment at the free end, as well as zero slope and zero deflection at the fixed end, namely,

$$\frac{d^3 u_z}{dx^3}\bigg|_{x=0} = 0 \quad \frac{d^2 u_z}{dx^2}\bigg|_{x=0} = 0 \quad \frac{d u_z}{dx}\bigg|_{x=l} = 0 \quad u_z(l) = 0 \tag{2.190}$$

By using them in Eqs. (2.185) through (2.188), a second-degree characteristic equation results whose roots are

$$\omega_1 = 3.53 \sqrt{\frac{EI_y}{ml^3}} \quad \omega_2 = 34.81 \sqrt{\frac{EI_y}{ml^3}} \tag{2.191}$$

While the first resonant frequency is very close to the exact one, the second one is larger than the corresponding exact one. A better approximation to this second resonant frequency can be obtained by using the shape functions

$$\varphi_1(x) = \frac{x}{l}\left(1 - \frac{x}{l}\right)^2 \quad \varphi_2(x) = \frac{x^2}{l^2}\left(1 - \frac{x}{l}\right)^2 \tag{2.192}$$

which produce:

$$\omega_1 = 4.468\sqrt{\frac{EI_y}{ml^3}} \qquad \omega_2 = 22.93\sqrt{\frac{EI_y}{ml^3}} \qquad (2.193)$$

It can be seen that while the second frequency is closer to the exact one, the natural frequency is a less precise approximation, compared to the one given by the first set of shape functions, Eqs. (2.189). A better approximation to both the first and second resonant frequencies can be produced by using three shape functions: the two in Eqs. (2.189) and the second Eqs. (2.192).

2.3.2 Circular rings

The circular ring can be used as a resonant microgyroscope, for instance, as discussed and implemented by Ayazi and Najafi.[9] This resonant microdevice is discussed in Chap. 5. A circular ring can vibrate radially, in bending, or in torsion or may have a combined torsional/bending mode, as discussed next.

The ring vibrates radially as shown in Fig. 2.36a. It can be assumed that the ring is only subject to axial deformations along its circumference during the radial vibrations.

In this case, it can be shown that the axial force that produces a radius change of u_r is

$$N = \frac{E A u_r}{R} \qquad (2.194)$$

The strain energy is therefore

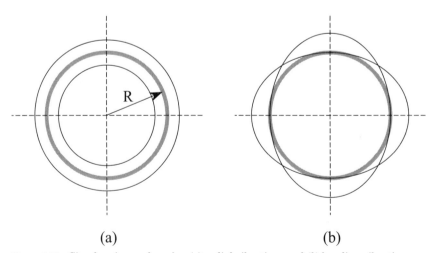

(a)   (b)

Figure 2.36 Circular ring undergoing (a) radial vibrations and (b) bending vibrations.

$$U = \frac{2 \pi R N^2}{2EA} = \frac{\pi E A u_r^2}{R} \qquad (2.195)$$

The elementary kinetic energy is

$$dT = \frac{\rho A R \dot{u}_r^2 \, d\varphi}{2} \qquad (2.196)$$

The total kinetic energy of the radially vibrating ring is calculated by integrating Eq. (2.196) over the whole circumference, which gives

$$T = \pi \rho A R \dot{u}_r^2 \qquad (2.197)$$

For conservative vibrations the sum of strain and kinetic energies is a constant, which means that its time derivative is zero:

$$\frac{d(T + U)}{dt} = 0 \qquad (2.198)$$

By combining Eqs. (2.195), (2.197), and (2.198) the following equation is obtained:

$$\ddot{u}_r + \omega_a^2 u_r = 0 \qquad (2.199)$$

where the resonant frequency is

$$\omega_a = \frac{1}{R}\sqrt{\frac{E}{\rho}} \qquad (2.200)$$

According to Timoshenko,[5] the first resonant frequency of a circular ring which vibrates torsionally (see Fig. 2.37) is

$$\omega_t = \frac{1}{R}\sqrt{\frac{E}{\rho} \frac{I_x}{I_t}} \qquad (2.201)$$

where I_x is the moment of inertia of the ring's cross section with respect to the x axis (which is the axis located in the plane of the ring, as shown in Fig. 2.37), whereas I_t is the torsional moment of inertia of the ring's

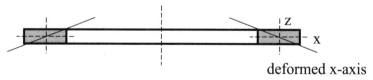

deformed x-axis

Figure 2.37   Circular ring undergoing free torsional vibrations.

cross section; both have been previously discussed when we analyzed straight-line members earlier in this chapter.

The circular ring can also undergo bending vibrations, as indicated in Fig. 2.36b. Timoshenko[5] suggests the following equation for resonant frequency calculations:

$$\omega_{b,i} = \frac{1}{R^2}\sqrt{\frac{i^2(i^2-1)}{i^2+1}\frac{EI_z}{\rho A}} \tag{2.202}$$

where i is the mode number. For $i = 1$, the result is a rigid body motion (no resonant frequency), and the first resonant frequency is obtained for $i = 2$ as

$$\omega_b = 6.726\sqrt{\frac{EI_z}{mR^3}} \tag{2.203}$$

For mixed torsional-bending vibrations the modal frequencies change to

$$\omega_{t-b,i} = \frac{1}{R^2}\sqrt{\frac{i^2(i^2-1)}{i^2+1+\mu}\frac{EI_z}{\rho A}} \tag{2.204}$$

where μ is Poisson's ratio. The first resonant frequency ($i = 2$) becomes

$$\omega_{t-b} = 15.04\sqrt{\frac{EI_z}{(5+\mu)mR^3}} \tag{2.205}$$

Example: Compare the resonant frequencies corresponding to axial, torsional, bending, and torsional-bending vibrations of a circular ring having very thin rectangular cross section of thickness t and width w ($t<<w$).

It can be shown that the axial and torsional resonant frequencies are related as

$$\omega_t = \frac{\omega_a}{2} \tag{2.206}$$

Also, by comparing Eqs. (2.203) and (2.205) it follows that

$$\frac{\omega_b}{\omega_{t-b}} = 0.447\sqrt{5+\mu} \tag{2.207}$$

For polysilicon where $\mu = 0.25$, Eq. (2.207) results in

$$\omega_b = 1.024\omega_{t-b} \tag{2.208}$$

which indicates that the bending and torsion-bending resonant frequencies are almost identical for very thin rectangular cross-section rings.

A comparison between the torsional and bending resonant frequencies [Eqs. (2.201) and (2.203)] leads to the relationship

$$\frac{\omega_t}{\omega_b} = 0.036\frac{R}{w} \qquad (2.209)$$

which shows that the torsional frequency is the smallest for designs where $R/w > 1/0.036 = 27.46$.

2.3.3 Thin plates and membranes

Thin plates and membranes are utilized as microresonators in fluidic applications, for instance, where bending resonant vibration of a thin member (usually clamped along its contour) realizes pumping of the fluid in or out of a cavity.

Thin plates (often referred to as Kirchhoff plates) are characterized by their small thickness compared to the other two (in-plane) dimensions; generally, a plate is considered thin when its thickness is less than one-twentieth of the smallest in-plane dimension. Such a plate is defined by its middle surface, and planes that are perpendicular to the original (undeformed) middle surface remain planar and perpendicular on the deformed middle surface. Another modeling feature/assumption of thin plates is that the middle surface does not stretch or compress under load. A thin plate can be defined by its deflection w (deformation out of the plane about the z direction), and this is a function of both x and y (see Fig. 2.38).

In characterizing the free bending response of a thin plate, the out-of-the-plane deformation w is of the form:

$$w(x, y, t) = w_0(x, y)\sin(\omega t) \qquad (2.210)$$

It can be shown (as demonstrated by Timoshenko,[5] for instance) that the bending resonant frequency is given by the Rayleigh procedure

Figure 2.38 Thin plate of constant thickness.

(outlined previously in this distributed-parameter section) and can be expressed as

$$\omega^2 = \frac{2}{\rho t} \frac{U_{max}}{\int_A w_0^2(x, y)\, dA} \tag{2.211}$$

where U_{max} is the maximum potential energy and is formulated as

$$U_{max} = \frac{D}{2} \int_A \left[\begin{array}{c} \left(\frac{\partial^2 w_0}{\partial x^2}\right)^2 + \left(\frac{\partial^2 w_0}{\partial y^2}\right)^2 + 2\mu\left(\frac{\partial^2 w_0}{\partial x^2}\right)\left(\frac{\partial^2 w_0}{\partial y^2}\right) \\ + 2(1 - \mu)\left(\frac{\partial^2 w_0}{\partial x\, \partial y}\right)^2 \end{array} \right]\, dA \tag{2.212}$$

The flexural rigidity D is defined as

$$D = \frac{Et^3}{12(1 - \mu^2)} \tag{2.213}$$

For a rectangular thin plate which is fixed (clamped on the contour), as sketched in Fig. 2.39, a suitable w_0 function needs to be used, that would satisfy the boundary conditions, namely, zero deflections and zero slopes (partial derivatives of deflection in terms of x or y) on the contour.

Such a function could be the following one:

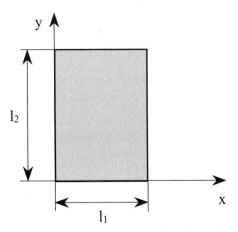

Figure 2.39 Top view of rectangular plate with dimensions and reference frame.

$$w_0(x, y) = a \frac{x^2}{l_1^2}\left(1 - \frac{x}{l_1}\right)^2 \frac{y^2}{l_2^2}\left(1 - \frac{y}{l_2}\right)^2 \qquad (2.214)$$

By taking w_0 (x, y) of Eq. (2.214) and substituting it into Eqs. (2.211) and (2.212), the following bending resonant frequency is obtained:

$$\omega^2 = \frac{6Et^2}{\rho(1 - \mu^2)} \frac{7(l_1^4 + l_2^4) + 4l_1^2 l_2^2}{l_1^4 l_2^4} \qquad (2.215)$$

A similar expression can be determined for a circular thin plate by using the generic formulation of Eqs. (2.211) and (2.212) and an appropriate w_0 function. Timoshenko[5] indicates the following equation:

$$\omega = \frac{10.21}{R^2} \sqrt{\frac{D}{\rho t}} \qquad (2.216)$$

While a thin plate is subjected to bending, a membrane (which can be structurally identical to a plate) is acted upon by a uniform stretching of its middle surface in a manner that would make negligible the small deflections occurring during vibration (Timoshenko[5]). A thin plate can be seen as the two-dimensional counterpart of a long beam, whereas the membrane is the two-dimensional correspondent of a bar subject to tension. We can find the first resonant frequency of a membrane by applying Rayleigh's procedure again and therefore by equating the maximum potential energy to the maximum kinetic energy. It can be shown (see Timoshenko,[5] for instance) that the resonant frequency of a generic membrane is calculated as

$$\omega^2 = \frac{s}{\rho t} \frac{\int_A [(\partial w_0/\partial x)^2 + (\partial w_0/\partial y)^2]\, dA}{\int_A w_0^2\, dA} \qquad (2.217)$$

where s represents a uniform tension per unit length of the boundary (measured in newtons per meter, for instance), and w_0 is the deflection of the membrane—it was defined in Eq. (2.210)—and should conveniently chosen to satisfy the boundary conditions of the membrane. Timoshenko[5] shows that the lowest resonant frequency of the rectangular membrane sketched in Fig. 2.39 is

$$\omega = 0.5 \sqrt{\frac{s}{\rho t}\left(\frac{1}{l_1^2} + \frac{1}{l_2^2}\right)} \qquad (2.218)$$

Similarly, the first resonant frequency of a circular membrane of radius R and thickness t is

$$\omega = \frac{2.404}{R} \sqrt{\frac{s}{\rho t}} \qquad (2.219)$$

References

1. W. C. Young and R. G. Budynas, *Roark's Formulas for Stress and Strain*, 7th ed., McGraw-Hill, New York, 2002.

2. N. Lobontiu, *Compliant Mechanisms: Design of Flexure Hinges*, CRC Press, Boca Raton, Fla., 2002.

3. N. Lobontiu and E. Garcia, *Mechanics of Microelectromechanical Systems*, Kluwer Academic Press, New York, 2004.

4. N. Lobontiu, E. Garcia, M. Hardau, and N. Bal, Stiffness characterization of corner-filleted flexure hinges, *Review of Scientific Instruments*, **75**(11), 2004, pp. 4896–4905.

5. S. Timoshenko, *Vibration Problems in Engineering*, D. Van Nostrand Company, New York, 1928.

6. W. T. Thomson, *Theory of Vibration with Applications*, 2nd ed., Prentice-Hall, Englewood Cliffs, N.J., 1981.

7. A. P. Boresi, R. J. Schmidt, and O. M. Sidebottom, *Advanced Mechanics of Materials*, 5th ed., Wiley, New York, 1993.

8. F. P. Beer, and E. R. Johnston, Jr., *Vector Mechanics for Engineers—Dynamics*, McGraw-Hill, Boston, 1997.

9. F. Ayazi and K. Najafi, A HARPSS polysilicon vibrating ring gyroscope, *Journal of Microelectromechanical Systems*, **10**(2), 2001, pp. 169–179.

3

Microhinges and Microcantilevers: Lumped-Parameter Modeling and Design

3.1 Introduction

This chapter focuses on microhinges and microcantilevers, specifically on determining their lumped-parameter stiffness and inertia properties to enable calculation of the corresponding resonant frequencies. The material presented here builds upon the developments of Chap. 2 as complex-shaped microhinges and microcantilevers can be fabricated by combining the basic shapes introduced in Chap. 2.

Microhinges are elastic regions of monolithically built MEMS that connect rigid parts and enable relative motions between these parts by either bending or torsion elastic deformation. A microhinge, which is connected to the substrate (ground) at one end and to a rigid body (mass) at the other end, is illustrated in Fig. 3.1.

From a modeling viewpoint, hinges in the lumped-parameter domain are considered as fixed-free members. Hinges in monolithic architectures have been proposed for macro-scale applications in various configurations such as right circular (Paros and Weisbord[1]) elliptical (Smith et al.[2]), or corner-filleted (Lobontiu et al.[3]). Other configurations including parabolic and hyperbolic, with two or multiple sensitive axes, have been studied by Lobontiu[4] from a compliance standpoint, whereas Lobontiu and Garcia[5] more recently provided the main elastic properties of a few hinge designs that can be implemented in MEMS applications.

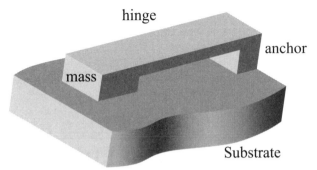

Figure 3.1 Microhinge with suspended mass.

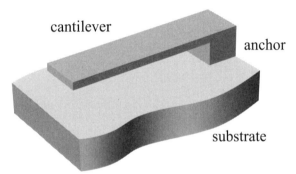

Figure 3.2 Suspended microcantilever.

Microcantilevers are physically fixed-free members, as shown in Fig. 3.2. The microcantilevers are employed in *atomic force microscopy* (AFM) for sub-Angstrom resolution reading and writing, as well as in micro- and nanotransduction applications such as detection of very small amounts of deposited substances. Other microcantilever applications are implemented in material property probing, cellular engineering, surface imaging, and metrology. The microcantilever can be actuated and monitored in the quasi-static regime or in the modal one by experimentally measuring the deflection or the slope (in the quasi-static regime) or the relevant resonant frequency (either the bending or the torsional ones). Examples of using microcantilevers for mass deposition purposes include the works of, to cite just a few, Raitieri et al.;[6] Ilic et al.;[7] Sato et al.;[8] Ilic, Yang, and Craighead;[9] Zalatudinov et al.;[10] or Britton et al.[11]. Atomic force microscopy applications have been studied, among many others, by King et al.;[12] Peterson et al.;[13] Morita, Wiesendanger, and Meyer;[14] van de Water and Molenaar;[15] and Ried et al.[16]. Other microcantilever geometries that are based on adding and subtracting circular and elliptical areas to and from rectangular regions are presented by Lobontiu and Garcia[17] and Garcia, Lobontiu, and Nam[18] who used an analytical model to evaluate both the static and the modal behavior of these members.

As Figs. 3.1 and 3.2 do indicate, microhinges and microcantilevers can be constructively and structurally identical, the only difference consisting in the boundary conditions: While microcantilevers are physically free at one end, microhinges are often connected at both ends to either the substrate or other rigid members of the micromechanism. Both flexible members are directly amenable to the same lumped-parameter modeling as fixed-free members (this is natural and direct for microcantilevers, but also valid for microhinges, which can be considered as fixed-free members). As a consequence, microhinges and microcantilevers are treated unitarily in this chapter.

Microhinges and microcantilevers are fully determined by means of lumped-parameter properties through 6 degrees of freedom associated to the free endpoint, namely, three translations (u_x, u_y, and u_z) and three rotations (θ_x, θ_y, and θ_z), as mentioned in Chap. 2 and shown in Fig. 2.1.

Determining the relevant stiffness properties of the various geometric configurations presented here often requires one to express compliances as a simpler calculation route. On occasion, compliances need to be determined in terms of reference frames that are not placed at one end of the member. This subsection addresses such designs by formulating the necessary quantitative rules. Several microcantilevers, including trapezoid, paddle, and circularly and elliptically filleted designs, are studied based on a generic approach which treats them as two-segment members to derive their lumped-parameter resonant frequencies. Circularly and elliptically filleted microhinges are also thoroughly presented by applying the same generic calculation tool.

Hollow microcantilevers such as rectangular and trapezoidal are further presented in this chapter. The resonant frequencies of multimorph (sandwiched) line members are also derived by focusing on two categories: the equal-length structures and the dissimilar-length ones. The chapter concludes by presenting the microcantilever arrays.

3.2   Compliance Transforms by Reference Frame Translation

Microhinges and microcantilevers can be designed as serial connections of elementary flexible segments which can be defined geometrically by a unique mathematical function (line, circle, or other curve), and which have been presented in Chap. 2 as basic units. In such cases a global reference frame is associated to all different segments making up the compound configuration. Different segments are, however, defined in

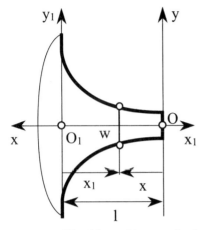

Figure 3.3 Microhinge with two end reference frames.

terms of compliances by means of local frames, which are translated from the global frame. Thus it becomes important to express compliances (and stiffnesses, by consequence) in reference frames that are translated from the original ones. Two types of problems relating to compliance transforms by reference frame translation are solved here in a generic fashion, namely, (1) relating the relevant compliances that are formulated in the two reference frames located at the member's ends and (2) expressing the compliances in a reference frame that is arbitrarily translated from the member's end in terms of the regular compliances.

3.2.1 Compliances in opposite-end reference frames

Figure 3.3 shows a microhinge, which has a variable width. This variable width can be expressed in terms of two reference frames, yOx (which is located at the thinner end) and $y_1O_1x_1$ (which is located at the opposite end). Compliances are usually calculated, particularly for microcantilevers which are thinner toward the free end and thicker toward the fixed root, in terms of the yOx frame. The necessity arises, however, to calculate compliances with respect to the other frame $y_1O_1x_1$, and therefore it would be useful to express the compliances in the $y_1O_1x_1$ frame in terms of the compliances in the yOx frame, which are usually available in the literature for a variety of geometric configurations.

As previously shown, three compliances define the bending about the y axis, namely, C_l (the linear direct bending compliance), C_r (the rotary direct bending compliance), and C_c (the cross-bending compliance), all

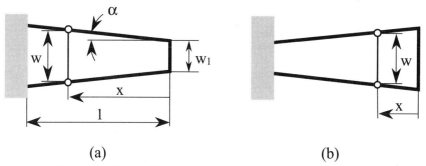

(a)   (b)

Figure 3.4 Trapezoid microcantilever: (*a*) direct configuration; (*b*) reversed configuration.

defined in Eqs. (2.27). If one denotes C_l', C_r', and C_c' as the linear, rotary, and cross compliances, respectively, of the same microhinge calculated now about the y_1 axis, and if the connection equations are considered

$$x = l - x_1 \quad dx = -dx_1 \tag{3.1}$$

then it can easily be demonstrated that the latter compliances are related to the former ones by means of the equations

$$C_l' = C_l - 2l\,C_c + l^2 C_r \quad C_r' = C_r \quad C_c' = -C_c + l\,C_r \tag{3.2}$$

In axial loading and torsion, the two compliances are identical, namely,

$$C_a' = C_a \quad C_t' = C_t \tag{3.3}$$

The prime superscript (') has been used to denote compliances taken with respect to the $x_1 O_1 y_1$ reference frame.

Example: We now analyze the trapezoid microcantilever shown in Fig. 3.4*a*, whose lumped-parameter resonant properties have been explicitly given in Chap. 2. It can be shown that its out-of-the-plane, bending-related compliances (linear, cross, and rotary) are

$$C_l = \frac{6l^3\,[(w_1 - w_2)(3w_1 - w_2) + 2w_1^2 \ln(w_2/w_1)]}{Et^3(w_2 - w_1)^3} \tag{3.4}$$

$$C_c = \frac{12l^2[w_2 - w_1 - w_1 \ln(w_2/w_1)]}{Et^3(w_2 - w_1)^2} \tag{3.5}$$

$$C_r = \frac{12l \ln(w_2/w_1)}{Et^3(w_2 - w_1)} \tag{3.6}$$

By applying now Eqs. (3.2), the compliances of the microcantilever shown in Fig. 3.4b are obtained (when the fixed and free ends interchange and the compliances are determined with respect to the new free end) as

$$C_l' = \frac{6l^3[(w_2 - w_1)(w_1 - 3w_2) + 2w_2^2 \ln(w_2/w_1)]}{Et^3(w_2 - w_1)^3} \tag{3.7}$$

$$C_c' = \frac{12l^2[w_1 - w_2 + w_2 \ln(w_2/w_1)]}{Et^3(w_2 - w_1)^2} \tag{3.8}$$

Equations (3.7) and (3.8) are also obtained when one is directly calculating the compliances C_l' and C_c' by applying the definition equations of linear direct and cross compliances of Chap. 2 and by starting the required integrations from the thicker end (assumed free).

3.2.2 Compliances in arbitrarily translated reference frames

At times, integrals that are of the form pertaining to the axial, torsion, and bending compliances need to be calculated with respect to reference frames that are translated at a specific distance from one end of the member. These compliances, as shown in the following, can be expressed in terms of the normally defined compliances. Figure 3.5 shows a generic microcantilever whose compliances, evaluated with respect to a translated frame $x_1 O_1 y_1$, need to be expressed in terms of the compliances calculated with respect to the reference frame xOy. The relationships between the two coordinates as well as between their corresponding differentials are

$$x = a + x_1 \qquad dx = dx_1 \tag{3.9}$$

It can be shown that the bending-related compliances calculated with respect to the translated frame $x_1 O_1 y_1$ can be expressed in terms of the normal ones that are formulated in the endpoint reference frame xOy as

$$C_l'' = C_l + 2aC_c + a^2 C_r \qquad C_r'' = C_r \qquad C_c'' = C_c + aC_r \tag{3.10}$$

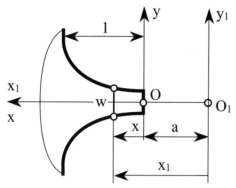

Figure 3.5 Microcantilever with arbitrarily translated reference frames.

Again, the relationship between axial and torsional compliances is, respectively,

$$C_a^{''} = C_a \qquad C_t^{''} = C_t \qquad (3.11)$$

The secondary superscript (") has been utilized to denote compliance taken in the $x_1 O_1 y_1$ (translated) reference frame.

Example: By taking the example of the microcantilever shown in Fig. 3.4a, its compliances taken with respect to a point situated at a distance

$$a = l_1 \qquad (3.12)$$

become

$$C_l^{''} = \frac{6l\{l(w_1 - w_2)[(3l + 4l_1)w_1 - (l + 4l_1)w_2] + 2[(l + l_1)w_1 - l_1 w_2]^2 \ln(w_2 / w_1)\}}{Et^3(w_2 - w_1)^3} \qquad (3.13)$$

$$C_c^{''} = \frac{12l\{l(w_2 - w_1) - [(l + l_1)w_1 - l_1 w_2] \ln (w_2/w_1)\}}{Et^3(w_2 - w_1)^2} \qquad (3.14)$$

3.3 Micromembers Formed of Two Compliant Segments

Microhinges and microcantilevers can be designed by serially connecting two different compliant segments, as sketched in Fig. 3.6, where the two portions have been represented by their center lines. The axial, torsion, and bending resonant frequencies are going to be

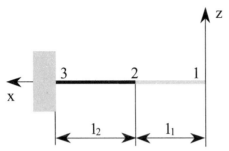

Figure 3.6 Two-segment microcantilever.

calculated in generic form with respect to the free end 1. The assumption is applied here that both segments have variable widths w_1 and w_2 but constant thicknesses t_1 and t_2.

The axial stiffness is

$$k_{a,e} = \frac{1}{C_a^{(1)} + C_a^{(2)}}$$ (3.15)

where $C_a^{(1)}$ and $C_a^{(2)}$ are the axial compliances of the two segments: (1) denotes the 1-2 portion and (2) denotes the 2-3 portion in the schematic representation of Fig. 3.6. The compliance $C_a^{(1)}$ is calculated with respect to point 1 whereas $C_a^{(2)}$ is determined with respect to point 2. The local frames of the two segments composing the member of Fig. 3.6 are placed at points 1 and 2, respectively. Equation (3.15) reduces to Eq. (2.45) when $l_1 = l_2 = l/2$, $w_2 = w_1$ (constant), and $t_1 = t_2$, which proves the validity of the former equation.

The lumped-parameter effective inertia fraction corresponding to the free axial vibrations is expressed as

$$m_{a,e} = \rho \left[t_1 \int_0^{l_1} w_1(x)[f_a^{(1)}(x)]^2 \, dx + t_2 \int_{l_1}^{l_1 + l_2} w_2(x)[f_a^{(2)}(x)]^2 dx \right]$$ (3.16)

The distribution functions $f_a^{(1)}$ and $f_a^{(2)}$ are assumed to be different for the two segments. When $l_1 = l_2 = l/2$, $w_2 = w_1$ (constant), and $t_1 = t_2$, Eq. (3.16) simplifies to Eq. (2.49), which characterizes a constant-cross-section microcantilever. The axially related resonant frequency can be calculated by means of Eqs. (3.15) and (3.16) and the definition Eq. (2.1).

The torsional stiffness formally resembles the axial one and is expressed as

$$k_{t,e} = \frac{1}{C_t^{(1)} + C_t^{(2)}}$$ (3.17)

and, again, its equation reduces to Eq. (2.51) which defines a constant-cross-section microcantilever in the case where $l_1 = l_2 = l/2$, $w_2 = w_1$ (constant), and $t_1 = t_2$.

It is interesting to check whether a relationship exists between the axial and torsional stiffnesses of a two-segment microcantilever in the case where the two segments have identical thicknesses $t_1 = t_2 = t$. Equation (3.15) can be rewritten as

$$k_{a,e} = \frac{Et}{\displaystyle\int_0^{l_1} dx/w_1(x) + \int_0^{l_2} dx/w_2(x)}$$ (3.18)

Similarly, Eq. (3.17) can be written in the form:

$$k_{t,e} = \frac{Gt^3}{3\left[\displaystyle\int_0^{l_1} dx/w_1(x) + \int_0^{l_2} dx/w_2(x)\right]}$$ (3.19)

Comparison of Eqs. (3.18) and (3.19) results in the following relationship:

$$k_{t,e} = \frac{Gt^2}{3E} k_{a,e}$$ (3.20)

Equation (3.20) is actually identical to Eq. (2.85), which applied for single-curve micocantilevers, as shown in Chap. 2. Equation (3.20) is not valid when the two segments have identical widths $w_1 = w_2 = w$ and different thicknesses; and as a consequence, the torsional stiffness will be explicitly calculated for the designs having this particular feature.

The lumped-parameter mechanical moment of inertia which corresponds to the free torsional vibrations of the serially compounded microcantilever of Fig. 3.6 is

$$J_{t,e} = \frac{\rho}{12}\left\{ t_1\int_0^{l_1} w_1(x)[f_a^{(1)}(x)]^2[w_1^2(x) + t_1^2]\,dx \right.$$

$$\left. + t_2\int_{l_1}^{l_1+l_2} w_2(x)[f_a^{(2)}(x)]^2[w_2^2(x) + t_2^2]\,dx \right\}$$ (3.21)

Equation (3.21) reduces to Eq. (2.55) when $l_1 = l_2 = l/2$, $w_2 = w_1$ (constant), and $t_1 = t_2$, when the serial microcantilever of Fig. 3.7 becomes a constant-cross-section member for the above-mentioned conditions.

In bending, it can be shown that the stiffness associated to the free end 1 is calculated as

$$k_{b,e} = \frac{1}{C_l^{(1)} + C_l^{(2)} + l_1^2 C_r^{(2)} + 2l_1 C_c^{(2)}} \tag{3.22}$$

Equation (3.22) is verified, too, because when $l_1 = l_2 = l/2$, $w_2 = w_1$ (constant), and $t_1 = t_2$, it reduces to Eq. (2.61), which characterizes a constant rectangular cross-section microcantilever of length l. The lumped effective mass is

$$m_{b,e} = \rho \left\{ t_1 \int_0^{l_1} w_1(x) \left[f_b^1(x) \right]^2 dx + t_2 \int_{l_1}^{l_1 + l_2} w_2(x) \left[f_b^{(2)}(x) \right]^2 dx \right\} \tag{3.23}$$

where $f_b^{(1)}$ and $f_b^{(2)}$ are the bending-related distribution functions, which are potentially different. Equation (2.66), which gives the effective inertia of a constant-cross-section microcantilever of length l, is retrieved from Eq. (3.23) when $l_1 = l_2 = l/2$, $w_2 = w_1$ (constant), and $t_1 = t_2$.

In the particular class of micromembers formed of two compliant segments, the two components are identical and placed in mirror, which means the resulting member has a symmetry axis, as sketched in Fig. 3.7. Purposely, the member of this structure has no axial symmetry about the x axis because this is not a prerequisite, although in the great majority of practical situations, microhinges and microcantilevers do also have axial symmetry. In such cases, the expression of bending stiffness, which is expressed generically in Eq. (3.22), can be simplified. The bending stiffness, which is calculated with respect to point 1 in Fig. 3.7, can be formulated by first expressing the linear direct bending compliance of the first (right-side) segment in terms of reversed frames—the first of Eqs. (3.2), namely,

$$C_l^{(1)\,\prime} = C_l^{(1)} - l C_c^{(1)} + \frac{l^2 C_r^{(1)}}{4} \tag{3.24}$$

and then applying the series-connection rule of Eq. (3.22). The bending stiffness of the flexible member of Fig. 3.7 becomes

$$k_{b,e} = \frac{1}{2(C_l^{(2)} + l^2 C_r^{(1)} / 4)} \tag{3.25}$$

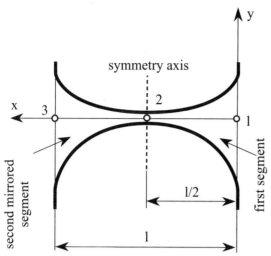

Figure 3.7 Micromember formed of two identical and mirrored compliant segments.

Equation (3.25) took into account the fact that the two halves composing the micromember of Fig. 3.7 are identical, and therefore have identical bending-related compliances (which add algebraically) when calculated by starting from point 2 with respect to both the (assumed) fixed point 1 (for the first segment) and 3 (for the second mirrored one).

Several examples are given next of microcantilevers and microhinges that are formed of two compliant segments by using the generic formulation presented in this section.

3.3.1 Paddle microcantilevers

A few microcantilever configurations are studied here that are formed as combinations of rectangular and/or trapezoid basic units of the types presented in Chap. 2. They are collectively named *paddle* microcantilevers, due to their shape; see Ilic, Yang, and Craighead.[9]

One such paddle configuration consists of two portions having different widths (w_2 at the fixed root is usually smaller than w_1, which is the width at the free end) and generally the same thickness t. Figure 3.8 shows the top view of a paddle microcantilever with its defining geometry.

The axial stiffness of the paddle microcantilever is

$$k_{a,e} = \frac{Etw_1w_2}{w_1l_2 + w_2l_1} \qquad (3.26)$$

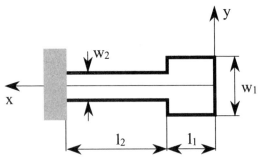

Figure 3.8 Top view and geometry of paddle microcantilever.

and it can be seen that when $l_1 = l_2$ and $w_1 = w_2$, Eq. (3.26) reduces to Eq. (2.45) which expresses the axial stiffness of a constant-cross-section cantilever. The lumped mass which is equivalent to the distributed inertia of the axially vibrating microrod is

$$m_{a,e} = \frac{\rho t[w_2 l_2^3 + w_1 l_1 (3l_2^2 + 3l_1 l_2 + l_1^2)]}{3(l_1 + l_2)^2} \qquad (3.27)$$

and this equation, too, simplifies to Eq. (2.49), yielding the effective mass of a constant rectangular cross-section microcantilever under the particular conditions $l_1 = l_2$ and $w_1 = w_2$.

The axial resonant frequency is

$$\omega_{a,e} = 1.73(l_1 + l_2) \sqrt{\frac{E(w_1 + w_2)}{\rho(w_1 l_2 + w_2 l_1)[w_2 l_2^3 + w_1 l_1 (3l_2^2 + 3l_1 l_2 + l_1^2)]}} \qquad (3.28)$$

The torsional stiffness is related to the axial stiffness according to Eq. (3.20). The mechanical moment of inertia, which is equivalent to the inertia corresponding to free torsional vibrations, is

$$J_{t,e} = \frac{\rho t[w_2 l_2^3 (w_2^2 + t^2) + w_1 l_1 (w_1^2 + t^2)(3l_2^2 + 3l_1 l_2 + l_1^2)]}{36(l_1 + l_2)^2} \qquad (3.29)$$

and this equation simplifies to Eq. (2.55) when the limit conditions $l_1 = l_2$ and $w_1 = w_2$ are satisfied. The torsional resonant frequency is found to be

$$w_{t,e} = 3.464(l_1 + l_2)t \sqrt{\frac{w_1 w_2 G}{\rho(w_1 l_2 + w_2 l_1)[w_2 l_2^3(w_2^2 + t^2)}{+w_1 l_1(w_1^2 + t^2)(3l_2^2 + 3l_1 l_2 + l_1^2)]}} \tag{3.30}$$

The bending stiffness is

$$k_{b,e} = \frac{Et^3 w_1 w_2}{4[w_2 l_1^3 + w_1 l_2(3l_1^2 + 3l_1 l_2 + l_2^2)]} \tag{3.31}$$

and when $l_1 = l_2$ and $w_1 = w_2$, it reduces to Eq. (2.61), which defines the linear direct bending stiffness of a constant rectangular cross-section microcantilever.

The lumped mass which is located at the free tip and is dynamically equivalent to the distributed inertia of the bending vibrating microcantilever is

$$m_{b,e} = \frac{\begin{aligned}\rho t[w_1 l_1^3(33l_1^4 + 231l_1^3 l_2 + 693l_1^2 l_2^2 \\ +1155 l_1 l_2^3 + 1155 l_2^4) + 63(10w_1 + w_2)l_1^2 l_2^5 \\ +7(20w_1 + 13w_2)l_1 l_2^6 + 33w_2 l_2^7]\end{aligned}}{140(l_1 + l_2)^6} \tag{3.32}$$

For $l_1 = l_2$ and $w_1 = w_2$, the mass fraction of Eq. (3.32) simplifies to Eq. (2.66), which gives the bending-related effective mass of a constant rectangular cross-section microcantilever. The bending-related resonant frequency is

$$w_{b,e} = \frac{5.92(l_1 + l_2)^3 t \sqrt{Ew_1 w_2} \left| \left\{ \begin{array}{c} \rho[w_2 l_1^3 + w_1 l_2 \\ \times (3l_1^2 + 3l_1 l_2 + l_2^2)] \end{array} \right\} \right|}{\sqrt{\begin{aligned} w_1 l_1^3(33l_1^4 + 231l_1^3 l_2 + 693l_1^2 l_2^2 + 1155 l_1 l_2^3 + 1155 l_2^4) \\ +63(10w_1 + w_2)l_1^2 l_2^5 + 7(20w_1 + 13w_2)l_1 l_2^6 + 33w_2 l_2^7 \end{aligned}}} \tag{3.33}$$

The microcantilever design of Fig. 3.9 resembles the configuration of Fig. 3.8, but it has constant width whereas the thicknesses of the two segments are different. The two segments are both flexible, and their location can be reversed (the thicker segment at the root and the thinner one at the free end).

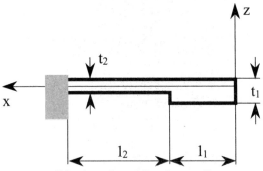

Figure 3.9 Side view of a paddle microcantilever.

By applying the composition rules that have been derived for two-segment microcantilevers, the axial stiffness is

$$k_{a,e} = \frac{E t_1 t_2 w}{l_1 t_2 + l_2 t_1} \tag{3.34}$$

The effective mass corresponding to free axial vibrations is

$$m_{a,e} = \frac{\rho [l_1 (l_1^2 + 3 l_1 l_2 + 3 l_2^2) t_1 + l_2^3 t_2] w}{3 (l_1 + l_2)^2} \tag{3.35}$$

The resulting axial resonant frequency is

$$\omega_{a,e} = 1.73 (l_1 + l_2) \sqrt{ \frac{E t_1 t_2}{\rho w (l_1 t_2 + l_2 t_1)[l_1 (l_1^2 + 3 l_1 l_2 + 3 l_2^2) t_1 + l_2^3 t_2]}} \tag{3.36}$$

The torsional stiffness is determined by means of Eq. (3.20) from the axial one. The effective torsional mechanical moment of inertia is

$$J_{t,e} = \frac{\rho w [l_1 t_1 (l_1^2 + 3 l_1 l_2 + 3 l_2^2)(t_1^2 + w^2) + l_2^3 t_2 (t_2^2 + w^2)]}{36 (l_1 + l_2)^2} \tag{3.37}$$

The torsional resonant frequency is

$$\omega_{t,e} = \frac{3.46 t_1 t_2 (l_1 + l_2) \sqrt{ \dfrac{G t_1 t_2}{\rho \, (l_1 t_2^3 + l_2 t_1^3)}}}{\sqrt{ l_1 t_1 (l_1^2 + 3 l_1 l_2 + 3 l_2^2)(t_1^2 + w^2) + l_2^3 t_2 (t_2^2 + w^2) }} \tag{3.38}$$

The bending stiffness is

$$k_{b,e} = \frac{Et_1^3 t_2^3 w}{4[l_2(3l_1^2 + 3l_1 l_2 + l_2^2)t_1^3 + l_1^3 t_2^3]} \qquad (3.39)$$

The effective bending mass is

$$m_{b,e} = \frac{\rho w\{l_1^3 t_1(33l_1^4 + 231l_1^3 l_2 + 693l_1^2 l_2^2 + 1155l_1 l_2^3 + 1155l_2^4) + l_2^5[33l_2^2 t_2 + 7l_1 l_2(20t_1 + 13t_2) + 63l_1^2(10t_1 + t_2)]\}}{140(l_1 + l_2)^6} \qquad (3.40)$$

The bending resonant frequency is

$$\omega_{b,e} = \frac{5.92 t_1 t_2(l_1 + l_2)^3 \sqrt{\dfrac{Et_1 t_2}{\rho\ [l_2(3l_1^2 + 3l_1 l_2 + l_2^2)t_1^3 + l_1^3 t_2^3]}}}{\sqrt{\begin{array}{l} l_1^3 t_1(33l_1^4 + 231l_1^3 l_2 + 693l_1^2 l_2^2 + 1155l_1 l_2^3 + 1155l_2^4) \\ + l_2^5[33l_2^2 t_2 + 7l_1 l_2(20t_1 + 13t_2) + 63l_1^2(10t_1 + t_2)]\end{array}}} \qquad (3.41)$$

All the lumped-parameter stiffness and inertia parameters corresponding to this microcantilever simplify to those of a constant rectangular cross-section design of length l, width w, and thickness t when $l_1 = l_2 = l/2$ and $t_1 = t_2$.

Example: Analyze how the bending resonant frequency of the paddle microcantilever sketched in Fig. 3.9 is influenced by the length and width ratios of the two segments. Consider the following numerical values: $\rho = 2300$ kg/m^3, $E = 150$ GPa, $l_1 = 500$ μm, and $t_1 = 3$ μm.

By using the relationships

$$\begin{aligned} l_2 &= c_l l_1 \\ t_2 &= c_t t_1 \end{aligned} \qquad (3.42)$$

and the given numerical values, the resonant bending frequency of Eq. (3.41) can be studied more closely, as illustrated in Fig. 3.10.

Figure 3.10 suggests that configurations with relatively shorter roots (shorter l_2 and therefore smaller values of c_l) and thicker ones (larger values of t_2 approaching t_1, which means higher c_t values) result in higher bending resonant frequencies.

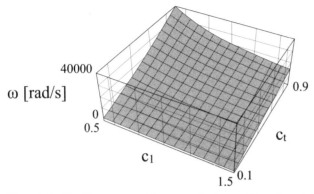

Figure 3.10 Bending resonant frequency in terms of length and thickness parameters.

Another paddle-type microcantilever comprising a rectangular portion at its root, connected to a trapezoid at its tip, is sketched in Fig. 3.11, which also gives the geometric parameters defining this configuration. The axial stiffness is

$$k_{a,e} = \frac{E t w_2 (w_1 - w_2)}{l_2(w_1 - w_2) + l_1 w_2 \ln(w_1 / w_2)} \tag{3.43}$$

The effective mass associated to the free axial vibrations is

$$m_{a,e} = \frac{\rho t [4 l_2^3 w_2 + 6 l_1 l_2^2 (w_1 + w_2) + 4 l_1^2 l_2 (2 w_1 + w_2) \\ + l_1^3 (3 w_1 + w_2)]}{12(l_1 + l_2)^2} \tag{3.44}$$

The corresponding resonant frequency is

$$\omega_{a,e} = \frac{3.46(l_1 + l_2) \sqrt{\dfrac{E w_2 (w_1 - w_2)}{\rho\, [l_2(w_1 - w_2) + l_1 w_2 \ln(w_1 / w_2)]}}}{\sqrt{\begin{array}{l} 4 l_2^3 w_2 + 6 l_1 l_2^2 (w_1 + w_2) + 4 l_1^2 l_2 (2 w_1 + w_2) \\ + l_1^3 (3 w_1 + w_2) \end{array}}} \tag{3.45}$$

The torsional stiffness, again, is connected to the axial one by means of Eq. (3.20). The torsion-related effective mechanical moment of inertia is

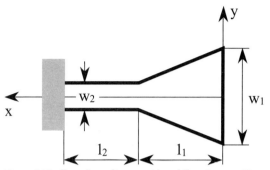

Figure 3.11   Top view of trapezoid paddle microcantilever.

$$J_{t,e} = \frac{\rho t \left\{ 20 l_2^3 w_2 (w_2^2 + t^2) + 15 l_1 l_2^2 (w_1 + w_2)(2t^2 + w_1^2 + w_2^2) \right.}{720(l_1 + l_2)^2}$$

$$+ l_1^3 [10 w_1^3 + 6 w_1^2 w_2 + 3 w_1 w_2^2 + w_2^3 + 5t^2(3 w_1 + w_2)]$$

$$\left. + 2 l_1^2 l_2 [10 t^2 (2 w_1 + w_2) + 3(4 w_1^3 + 3 w_1^2 w_2 + 2 w_1 w_2^2 + w_2^3)] \right\}$$

(3.46)

The torsional resonant frequency is too complex and is not calculated here, but it can simply be found by means of Eq. (3.46).
   The bending stiffness is

$$k_{b,e} = \frac{E t^3 w_2 (w_1 - w_2)^3}{2 \left\{ (w_1 - w_2) [2(w_1 - w_2)^2 l_2 (3 l_1^2 + 3 l_1 l_2 + l_2^2) + 3 l_1^3 w_2 (w_1 - 3 w_2)] + 6 l_1^3 w_2^3 \ln(w_1 / w_2) \right\}}$$

(3.47)

The effective bending mass is

$$m_{b,e} = \frac{\rho t [264 l_2^7 w_2 + 56 l_1^6 l_2 (26 w_1 + 7 w_2) + 56 l_1 l_2^6 (10 w_1 + 23 w_2)}{1120(l_1 + l_2)^6}$$

$$+ 56 l_1^2 l_2^5 (50 w_1 + 49 w_2) + 70 l_1^3 l_2^4 (83 w_1 + 49 w_2)$$

$$+ 56 l_1^4 l_2^3 (116 w_1 + 49 w_2) + 28 l_1^5 l_2^2 (149 w_1 + 49 w_2)$$

$$+ l_1^7 (215 w_1 + 49 w_2)]$$

(3.48)

Again, the bending resonant frequency is found by means of Eqs. (3.47) and (3.48) and is not included here. All the lumped-parameter stiffnesses and inertia fractions were checked by reformulating them for

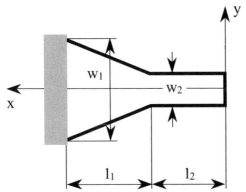

Figure 3.12   Top view of reversed trapezoid-root microcantilever.

$l_1 = l_2 = l/2$ and $w_1 = w_2$ and the corresponding properties characterizing a constant rectangular cross-section microcantilever of length l, thickness t, and width w have been obtained.

Example: Compare the bending stiffness, effective mass, and resonant frequency of the microcantilever sketched in Fig. 3.11 to the similar lumped-parameter amounts of the reversed microcantilever of Fig. 3.12.

The bending stiffness of this microcantilever is calculated by applying the series connection rule given in Eq. (3.22). The direct linear bending stiffness of the constant-cross-section segment located at the free end is given in Eq. (2.61), and the corresponding compliance is its algebraic inverse. The bending compliances (direct linear, direct rotary, and cross) of the root trapezoid segment are expressed in Eqs. (3.4) through (3.6), where w_2 has to be used instead of w_1 (and vice versa), and l_1 instead of l. The following bending ratio can be formulated:

$$rk_b = \frac{k_{b,e}^{\text{rev}}}{k_{b,e}} \tag{3.49}$$

where the superscript "rev" (for reversed) indicates the bending stiffness of the microcantilever of Fig. 3.12 and $k_{b,e}$ is the bending stiffness of the configuration shown in Fig. 3.11. The stiffness ratio of Eq. (3.49) can be expressed in terms of the following nondimensional parameters:

$$c_w = \frac{w_2}{w_1} \qquad c_l = \frac{l_2}{l_1} \tag{3.50}$$

This ratio is plotted in Fig. 3.13 which indicates that the configuration of Fig. 3.12 can be approximately 6 times stiffer in bending than that of Fig. 3.11. A similar comparison can be made in terms of the effective mass corresponding to bending. The mass ratio

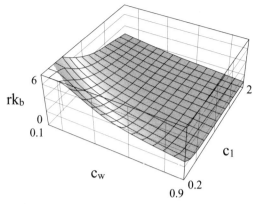

Figure 3.13 Bending stiffness comparison between the microcantilever of Fig. 3.11 and the reversed configuration of Fig. 3.12.

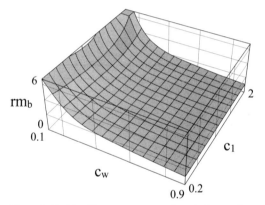

Figure 3.14 Bending mass comparison between the microcantilever of Fig. 3.11 and the reversed configuration of Fig. 3.12.

$$rm_b = \frac{m_{b,e}}{m_{b,e}^{rev}}$$ (3.51)

is now plotted in Fig. 3.14 in terms of the nondimensional parameters of Eq. (3.50).

As Fig. 3.14 indicates, the effective mass of the reversed microcantilever of Fig. 3.12 is smaller than that of the parent microcantilever of Fig. 3.11, and this is understandable, as the configuration with more mass toward the free end—that of Fig. 3.11—will generate more effective inertia at the same point. Both Figs. 3.12 and 3.13 show that the lumped-parameter stiffness and mass of the reversed microcantilever approach the corresponding amounts that characterize the parent configuration when the widths are approximately equal (large values of c_w) and when the length of the

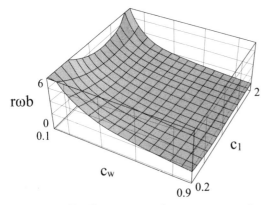

$r\omega b$

Figure 3.15 Bending resonant frequency comparison between the microcantilever of Fig. 3.11 and the reversed configuration of Fig. 3.12.

constant-cross-section segment is larger than that of the trapezoid (for values of c_l that are larger than 1).

It can be shown by using Eqs. (3.49) and (3.51) that the bending resonant frequency ratio can be expressed as

$$r\omega_b = \frac{\omega_{b,e}^{rev}}{\omega_{b,e}} = \sqrt{rk_b rk_m} \tag{3.52}$$

and Fig. 3.15 is the three-dimensional plot of this ratio.

The trend displayed by both the stiffness and the mass ratios is also followed by the resonant frequency ratio, which is plotted in Fig. 3.15, as this ratio is the square root of the product of the stiffness and mass ratios defined in Eqs. (3.49) and (3.51).

The paddle microcantilever of Fig. 3.16 is similar to the one previously studied, and it combines a rectangular portion at the root with a trapezoid at its free end. The difference consists in the trapezoid unit having its thickness variable. The two basic units have the same width w.

The axial stiffness is

$$k_{a,e} = \frac{Et_2(t_1 - t_2)w}{l_2(t_1 - t_2) + l_1 t_2 \ln(t_1 / t_2)} \tag{3.53}$$

The effective axial mass is

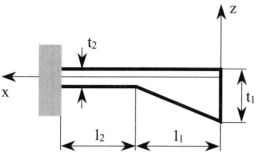

Figure 3.16 Side view of trapezoid paddle microcantilever.

$$m_{a,e} = \frac{\rho w[4l_2^3 t_2 + 6l_1 l_2^2(t_1 + t_2) + 4l_1^2 l_2(t_1 + 2t_2) + l_1^3(t_1 + 3t_2)]}{12(l_1 + l_2)^2} \tag{3.54}$$

The axial resonant frequency is

$$\omega_{a,e} = \frac{3.46(l_1 + l_2)\sqrt{\dfrac{Et_2(t_1 - t_2)}{\rho[l_2(t_1 - t_2) + l_1 t_2 \ln(t_1/t_2)]}}}{\sqrt{4l_2^3 t_2 + 6l_1 l_2^2(t_1 + t_2) + 4l_1^2 l_2(t_1 + 2t_2) + l_1^3(t_1 + 3t_2)}} \tag{3.55}$$

The torsional stiffness is computed by inverting the torsional compliance, which is calculated according to its definition in Eq. (2.26), and it is

$$k_{t,e} = \frac{2Gt_1^2 t_2^3 w}{3[2l_2 t_1^2 + l_1 t_2(t_1 + t_2)]} \tag{3.56}$$

The effective torsional mechanical moment of inertia is

$$J_{t,e} = \frac{\rho w\left\{20l_2^3 t_2(w^2 + t_2^2) + 15l_1 l_2^2(t_1 + t_2)(2w^2 + t_1^2 + t_2^2) + l_1^3[10t_2^3 + 6t_1 t_2^2 + 3t_1^2 t_2 + t_1^3 + 5w^2(t_1 + 3t_2)] + 2l_1^2 l_2[10w^2(t_1 + 2t_2) + 3(4t_2^3 + 3t_1 t_2^2 + 2t_1^2 t_2 + t_1^3)]\right\}}{720(l_1 + l_2)^2} \tag{3.57}$$

The torsional resonant frequency can now be calculated by means of the lumped-parameter stiffness [Eq. (3.56)] and mechanical moment of inertia [Eq. (3.57)].

The bending stiffness is

$$
k_{b,e} = \cfrac{Ewt_1^2 t_2^3 (t_1 - t_2)^3}{2 \big/ \big\{ 2t_1^2 (t_1 - t_2)^3 l_2 [(3l_1(l_1 + l_2) + l_2^2)] + 3l_1^3 t_2^3 [(3t_1 - t_2)(t_2 - t_1) + 2t_1^2 \ln(t_1 / t_2)] \big\}}
\tag{3.58}
$$

The bending effective mass is

$$
m_{b,e} = \cfrac{\rho w [264 l_2^7 t_2 + 56 l_1 l_2^6 (10t_1 + 23t_2) + 56 l_1^6 l_2 (7t_1 + 26t_2)}{1120(l_1 + l_2)^6}
$$

$$
\begin{aligned}
&+ 56 l_1^2 l_2^5 (40t_1 + 59t_2) + 70 l_1^3 l_2^4 (49t_1 + 83t_2) \\
&+ 56 l_1^4 l_2^3 (49t_1 + 116t_2) + 28 l_1^5 l_2^2 (49t_1 + 149t_2) \\
&+ l_1^7 (49t_1 + 215t_2)]
\end{aligned}
\tag{3.59}
$$

The bending resonant frequency is determined by combining Eqs. (3.58) and (3.59) and is not included here.

The following particular conditions have been used: $l_1 = l_2 = l/2$, $t_1 = t_2$ in Eqs. (3.53) through (3.59), which transformed the microcantilever of Fig. 3.16 to a constant rectangular cross-section microcantilever of length l, thickness t, and width w. Indeed, the corresponding Eqs. (2.45), (2.49), (2.51), (2.55), (2.61), and (2.66) were used, which define the constant rectangular cross-section microcantilever.

3.3.2 Filleted microcantilevers

Another possible combination of the basic units of Chap. 2 is studied here, namely, by realizing compound cantilevers through the mixing of circular and elliptic filleted units, either by themselves or with constant rectangular cross-section segments.

Long, circularly filleted microcantilevers. Figure 3.17 shows the top view with the main geometric parameters defining a long, circularly filleted microcantilever which consists of a constant rectangular area at the free end connected to a circularly filleted area at the fixed root. The lumped-parameter stiffness and inertia will again be determined in view of calculating the resonant frequencies of this micromember that

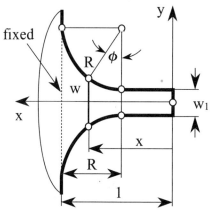

Figure 3.17 Geometry of long, circularly filleted microcantilever.

are associated with axial, torsional, and bending free vibrations. A similar treatment can also be found in Lobontiu and Garcia.[17]

In free axial vibrations the lumped-parameter stiffness which is associated to applying a force at the free end about the x direction and finding the corresponding x deformation is

$$
k_{a,e} = \frac{Et}{\begin{aligned}&(l - R)/w_1 + (2R + w_1)/\sqrt{w_1(4R + w_1)} \\ &\times \ \arctan\sqrt{1 + 4R/w_1} - \pi/4\end{aligned}}
$$

(3.60)

Equation (3.60) reduces to Eq. (2.45), which yields the axial stiffness of a constant-cross-section microcantilever when $R \to 0$. Moreover, as Fig. 3.17 indicates, the long, circularly filleted microcantilever transforms to a circularly filleted configuration, of the type displayed in Fig. 2.30 when $l \to R$, and this should be reflected in the corresponding mathematics. Indeed, when $l \to R$, Eq. (3.60) matches Eq. (2.130), which gives the axial stiffness of a circularly filleted microcantilever. The lumped-parameter mass corresponding to free axial vibrations of the microcantilever of Fig. 3.17 is

$$
m_{a,e} = \rho t \left(\frac{0.036R^4}{l^2} + \frac{lw_1}{3} \right)
$$

(3.61)

When $R \to 0$, Eq. (3.61) simplifies to Eq. (2.49), which yields the effective mass corresponding to the free axial vibrations of a constant rectangular cross-section microcantilever. Similarly, when $l \to R$, Eq. (3.61) simplifies to Eq. (2.131) which defines the effective axial mass of a circularly filleted microcantilever. The resonant frequency of the free axial vibration of the microcantilever shown in Fig. 3.17 is

$$\omega_{a,e} = \frac{\sqrt{E/\{\rho(0.036R^4/l^2 + lw_1/3)}}{\sqrt{\dfrac{(l-R)/w_1 + (2R+w_1)}{/\sqrt{w_1(4R+w_1)}\arctan\sqrt{1+4R/w_1} - \pi/4}}} \qquad (3.62)$$

In free torsional vibrations the lumped-parameter stiffness is expressed by means of Eq. (3.20) in terms of the axial stiffness given in Eq. (3.60). The lumped torsional mechanical moment of inertia is

$$J_{t,e} = \frac{\begin{array}{c}\rho t[0.832R^6 + 4.567R^5 w_1 + 4.381R^4(3w_1^2 + t^2) \\ + 40w_1 l^3(w_1^2 + t^2)]\end{array}}{1440 l^2} \qquad (3.63)$$

For $R \to 0$, Eq. (3.63) reduces to Eq. (2.55), which expresses the lumped-parameter inertia corresponding to torsional vibrations of a constant rectangular cross-section microcantilever. Also, when $l \to R$, Eq. (3.63) changes to Eq. (2.133), which yields the effective torsional mechanical moment of inertia of a circularly filleted microcantilever. The torsion-related resonant frequency of the microcantilever sketched in Fig. 3.17 is

$$\omega_{t,e} = \frac{21.91 lt \sqrt{\begin{array}{c}G/[0.832R^6 + 4.567R^5 w_1 \\ + 4.381R^4(3w_1^2 + t^2) + 40w_1 l^3(w_1^2 + t^2)]\end{array}}}{\sqrt{\dfrac{[(l-R)/w_1 + (2R+w_1)}{/\sqrt{w_1(4R+w_1)}\arctan\sqrt{1+4R/w_1} - \pi/4]}}} \qquad (3.64)$$

The lumped-parameter stiffness in bending is

$$k_{b,e} = \cfrac{Et^3}{\begin{aligned}&4(l-R)^3\big/w_1 + 3(l-R)^2[4(2R+w_1)\big/\sqrt{w_1(4b+w_1)} - \pi]\\ &+ 0.75[14.283R^2 + 16.566Rw_1 + 3.14w_1^2\\ &- 4(2R+w_1)\sqrt{w_1(4R+w_1)}\arctan\sqrt{1+4R/w_1}]\\ &+ 6(l-R)[(2R+w_1)\ln(1+2R/w_1) - 2R]\end{aligned}}$$

(3.65)

Equation (3.65) reduces to Eq. (2.61) when $R \to 0$, which demonstrates that the long, circularly filleted configuration of Fig. 3.17 transforms to a constant rectangular cross-section microcantilever. Also, when $l \to R$, the same Eq. (3.65) changes to Eq. (2.135), which defines the bending stiffness of a circularly filleted configuration. The lumped-parameter mass, which is equivalent to the distributed inertia of the microcantilever of Fig. 3.17 during free bending vibrations, is

$$m_{b,e} = \frac{\rho t[(1208.75l^2 - 498.31lR + 54.959R^2)R^6 + 12.672l^7 w_1]}{53.760l^6} \tag{3.66}$$

Equation (3.66) simplifies to Eq. (2.66) when $R \to 0$, as expected, whereas for $l \to R$ the same equation simplifies to Eq. (2.136), which gives the effective bending mass of a circularly filleted microcantilever. The resonant frequency corresponding to the free bending vibrations is

$$\omega_{b,e} = \cfrac{66.933l^3 t\sqrt{\begin{aligned}&E\big/\{\rho[(1208.75l^2 - 498.31lR + 54.959R^2)R^6\\ &+ 12672l^7 w_1]\end{aligned}}}{\sqrt{\begin{aligned}&4(l-R)^3\big/w_1 + 3(l-R)^2[4(2R+w_1)\\ &\big/\sqrt{w_1(4b+w_1)} - \pi] + 0.75[14.283R^2 + 16.566Rw_1\\ &+ 3.14w_1^2 - 4(2R+w_1)\sqrt{w_1(4R+w_1)}\arctan\sqrt{1+4R/w_1}]\\ &+ 6(l-R)[(2R+w_1)\ln(1+2R/w_1) - 2R]\end{aligned}}}$$

(3.67)

Example: Compare the resonant frequencies in bending and torsion of a long, circularly filleted microcantilever, which is defined by $l = 500$ μm and $t = 1$ μm when Poisson's ratio is $\mu = 0.25$.
The following nondimensional parameters are considered:

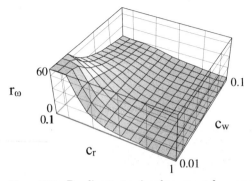

Figure 3.18 Bending-to-torsional resonant frequency ratio.

$$c_r = \frac{R}{l} \qquad c_w = \frac{w_1}{l} \tag{3.68}$$

By using these equations and the numerical values given here, the following resonant frequency ratio is formulated:

$$r_\omega = \frac{\omega_{b,e}}{\omega_{t,e}} \tag{3.69}$$

Where the bending and torsional resonant frequencies are given in Eqs. (3.67) and (3.64), the plot in Fig. 3.18 is obtained.

Figure 3.18 indicates that the bending frequency is higher than the torsional one for the numerical values of this example and for smaller values of the vaiables. As the fillet radius approaches the total microcantilever length and the width increases, the two resonant frequencies become comparable (their ratio approaches unity), and the bending resonant frequency becomes smaller than the torsional one, as illustrated in the same figure.

Long elliptically filleted microcantilevers. Another corner-filleted configuration is the elliptical one, as sketched in Fig. 3.19, which is formed of an elliptically filleted unit at the root and a constant rectangular cross-section unit at the end. This configuration is also presented in Lobontiu and Garcia.[5, 17] The two filleted parts are quarter ellipses defined by the semiaxes a and b.

By using the serial connection rule introduced in this chapter, the axial stiffness is expressed as

$$k_{a,e} = \frac{Et}{(l-a)\big/w_1 + a[(2b+w_1)\big/\sqrt{w_1(4b+w_1)}}$$
$$\times \arctan\sqrt{1+4b\big/w_1} - \pi\big/4]\big/b \tag{3.70}$$

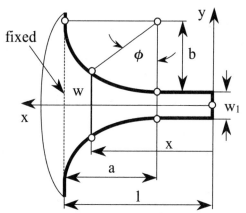

Figure 3.19 Elliptical corner-filleted microcantilever.

When $a \to 0$, this equation reduces to Eq. (2.45), which corresponds to a constant-cross-section cantilever of length l. Similarly, when $a \to R$ and $b \to R$, Eq. (3.70) changes to Eq. (3.60), which describes a long, circularly filleted microcantilever.

The effective mass in free axial vibrations is

$$m_{a,e} = \rho t \left(\frac{0.036a^3 b}{l^2} + \frac{l w_1}{3} \right) \tag{3.71}$$

which again transforms to Eq. (2.49) for $a \to 0$ (constant rectangular cross-section microcantilever) and to Eq. (3.61) when $a \to R$ and $b \to R$ (long, circularly filleted microcantilever). The axial resonant frequency is

$$\omega_{a,e} = \frac{\sqrt{E / [\rho(0.036a^3 b / l^2 + l w_1 / 3)]}}{\sqrt{\begin{array}{c} (l - a) / w_1 + a[(2b + w_1) / \sqrt{w_1(4b + w_1)} \\ \times \arctan\sqrt{1 + 4b / w_1} - \pi / 4] / b \end{array}}} \tag{3.72}$$

Again, the torsional stiffness can be found from the axial one by means of Eq. (3.20). The effective torsional moment of inertia is

$$J_{t,e} = \frac{\rho t \{a^3 b[0.832b^2 + 4.567b w_1 + 4.381(3w_1^2 + t^2)] + 40w_1 l^3(w_1^2 + t^2)\}}{1440 l^2} \tag{3.73}$$

For $a \to 0$, Eq. (3.73) simplifies to Eq. (2.55), which is the effective torsional mechanical moment of inertia of a constant-cross-section member, whereas for $a \to R$ and $b \to R$, the long, elliptically filleted cantilever becomes a circularly filleted one, and indeed, Eq. (3.73) changes to Eq. (3.63), as it should. The lumped-parameter torsional resonant frequency is found by combining the corresponding stiffness and inertia fractions, and its equation is

$$\omega_{t,e} = \frac{21.91lt\sqrt{\begin{array}{l}(G/\rho)/\left\{a^3 b[0.832b^2 + 4.56bRw_1\right.\\ \left.+4.381(3w_1^2 + t^2)] + 40w_1 l^3(w_1^2 + t^2)\right\}\end{array}}}{\sqrt{\begin{array}{l}(l-a)/w_1 + a[(2b + w_1)\\ /\sqrt{w_1(4b + w_1)}\ \arctan\sqrt{1 + 4b/w_1} - \pi/4]/b\end{array}}} \tag{3.74}$$

The bending stiffness for a long, elliptically filleted cantilever is found by means of the series connection rule of Eq. (3.22) as

$$k_{b,e} = \frac{Et^3}{\begin{array}{l}4(l-a)^3/w_1 + 3a(l-a)^2[4(2b + w_1)\\ /\sqrt{w_1(4b + w_1)} - \pi]/b + 0.75a^3[14.283b^2\\ +16.566bw_1 + 3.14w_1^2\\ -4(2b + w_1)\sqrt{w_1(4b + w_1)}\ \arctan\sqrt{1 + 4b/w_1}]/b^3\\ +6a^2(l-a)[(2b + w_1)\ln(1 + 2b/w_1) - 2b]/b^2\end{array}} \tag{3.75}$$

When any of the two segments composing the elliptically filleted microcantilever of Fig. 3.19 is relatively short, and shearing effects need be taken into account, the shearing-affected compliances have to be used in formulating the corresponding bending stiffness. The effective inertia fraction associated with the free bending vibrations is

$$m_{b,e} = \frac{\rho t[(1208.75l^2 - 498.31la + 54.959a^2)a^5 b + 12,672l^7 w_1]}{53,760l^6} \tag{3.76}$$

Again, when shearing effects are important, the shearing-affected distribution function has to be utilized, instead of the normal one, and an effective mass, different from that of Eq. (3.76) will be obtained. The same limit calculations that have been performed before, namely, by considering that $a \to 0$ and $a \to R$ and $b \to R$, respectively, have been performed on Eqs. (3.75) and (3.76). The expected equations corresponding to constant rectangular cross-section and long, circularly filleted microcantilevers have been obtained. The bending-related resonant frequency can be found by combining Eqs. (3.75) and (3.76), and its equation is

$$
\omega_{b,e} = \cfrac{66.933 l^3 t \sqrt{\begin{array}{c} E/\rho[(1208.75l^2 - 498.31la + 54.959a^2)a^5 b \\ +12{,}672l^7 w_1] \end{array}}}{\sqrt{\begin{array}{c} 4(l-a)^3/w_1 + 3a(l-a)^2[4(2b+w_1)/\sqrt{w_1(4b+w_1)} \\ -\pi]/b + 0.75a^3[14.283b^2 + 16.566bw_1 + 3.14w_1^2 \\ -4(2b+w_1)\sqrt{w_1(4b+w_1)}\arctan\sqrt{1 + 4b/w_1}]/b^3 \\ +6a^2(l-a)[(2b+w_1)\ln(1 + 2b/w_1) - 2b]/b^2 \end{array}}}
$$

(3.77)

Example: A long, elliptically filleted microcantilever is defined by the geometry parameters $l = 300$ µm, $w_1 = 10$ µm, and $t = 2$ µm and material properties $E = 150$ GPa and $\rho = 2300$ kg/m³. Study the bending resonant frequency of this microcantilever in terms of the semiaxis lengths a and b.

When the lengths of the semiaxes are allowed to range within the interval [10 µm, 300 µm], the bending resonant frequency of Eq. (3.77) varies according to the three-dimensional plot of Fig. 3.20.

As expected, the resonant frequency increases with both a and b increasing, as illustrated in Fig. 3.20, which indicates that stiffness dominates over

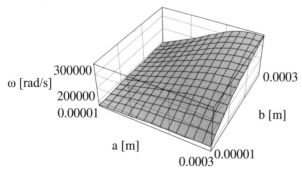

Figure 3.20   Bending resonant frequency of long, elliptically filleted microcantilever in terms of semiaxis lengths a and b.

inertia in this trend. It can also be seen that for relatively small values of a (the "length" of the fillet) the increase in resonant frequency is somewhat limited, irrespective of the increase in b; conversely, this is also valid for relatively small values of b.

3.3.3 Filleted microhinges

By combining again filleted units (either circular or elliptical) and constant rectangular cross-section units, compound members can be obtained, which have a center of symmetry and which can be utilized as flexible connectors, called microhinges. The microhinges are generally active through bending and/or torsion, and therefore these two vibrational modes together with the axial one are analyzed. Three configurations are studied in this section, namely, the right circular, the right elliptical, and the circular corner-filleted microhinges.

Right circular microhinges. A very popular configuration is the right circular microhinge, sketched in Fig. 3.21, which is formed of two right circularly filleted units, such as those studied in Chap. 2.

This type of compliant connector was probably the first one to receive attention as early as the 1960s when Paros and Weisbord[1] derived the analytical compliances pertaining to bending (both about the sensitive axis — the y axis here and the z axis) and axial loading. Not only is this configuration utilized as a microhinge, but also it can be part of another two-segment microcantilever configuration — the circularly notched one — which is analyzed later in this chapter.

By considering that the microhinge of Fig. 3.21 is composed of two identical and mirrored circularly filleted designs, such as the one

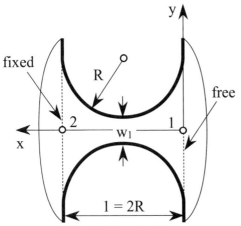

Figure 3.21 Right circular microhinge.

studied in Chap. 2, it is possible to determine its axial, torsional, and bending resonant frequencies by applying the algorithms developed for two-segment configurations. It can be shown that the axial stiffness (the axial compliance is given in Lobontiu,[4] for instance), which is connected to the free end 1 being derived with respect to the fixed end 2 and is found by using the series connection Eq. (3.22), is

$$k_{a,e} = \frac{Et}{2[(2R + w_1)/\sqrt{w_1(R + w_1)}\arctan\sqrt{1 + 4R/w_1} - \pi/4]}$$ (3.78)

The effective axial mass is

$$m_{a,e} = \rho t R(0.352R + 0.667w_1)$$ (3.79)

The corresponding axial resonant frequency is

$$\omega_{a,e} = \frac{0.707\sqrt{E/[\rho R(0.352R + 0.667w_1)]}}{\sqrt{(2R + w_1)/\sqrt{w_1(R + w_1)}\arctan\sqrt{1+4R/w_1} - \pi/4}}$$ (3.80)

The torsional stiffness is calculated by means of Eq. (3.20) in terms of the axial stiffness expressed in Eq. (3.78). The effective torsional mechanical moment of inertia is

$$J_{t,e} = \frac{\rho t R}{12[0.403R^3 + 1.019R^2w_1 + 0.667w_1(w_1^2 + t^2) + 0.352R(3w_1^2 + t^2)]}$$ (3.81)

The torsion-related resonant frequency is

$$\omega_{t,e} = \frac{1.414t\sqrt{\dfrac{G/\{\rho R[(2R + w_1)}{/\sqrt{w_1(R + w_1)}\arctan\sqrt{1 + 4R/w_1} - \pi/4]\}}}}{\sqrt{0.403R^3 + 1.019R^2w_1 + 0.667w_1(w_1^2 + t^2) + 0.352R(3w_1^2 + t^2)}}$$ (3.82)

The direct linear (out-of-the-plane) bending stiffness of a long (Euler-Bernoulli) configuration is

$$k_{b,e} = \cfrac{Et^3}{\begin{aligned}&2.572R^2 + 24.849Rw_1 + 4.713w_1^2\\&+ 6(2R + w_1)(4R^2 - 4Rw_1 - w_1^2)\\&\big|\sqrt{w_1(4R + w_1)}\arctan\sqrt{1 + 4R/w_1}\end{aligned}} \qquad (3.83)$$

The effective mass associated with the free bending vibrations is

$$m_{b,e} = \rho Rt(0.308R + 0.47w_1) \qquad (3.84)$$

The resulting bending-related resonant frequency is

$$\omega_{b,e} = \cfrac{t\sqrt{\dfrac{E}{\rho R(0.308R + 0.471w_1)}}}{\sqrt{\begin{aligned}&2.572R^2 + 24.849Rw_1 + 4.713w_1^2\\&+6(2R + w_1)(4R^2 - 4Rw_1 - w_1^2)\\&\big|\sqrt{w_1(4R + w_1)}\arctan\sqrt{1 + 4R/w_1}\end{aligned}}} \qquad (3.85)$$

For relatively short configurations, the Timoshenko model needs to be utilized, and the resulting bending stiffness is

$$k_{b,e}^{sh} = \cfrac{2EGt^3}{\begin{aligned}&3G(1.717R^2 - 16.567Rw_1 + 3.142w_1^2) - \pi\kappa Et^2\\&+4(2R + w_1)[\kappa Et^2 + 3G(4R^2 - 4Rw_1 - w_1^2)]\\&\big|\sqrt{w_1(4R + w_1)}\arctan\sqrt{1 + 4R/w_1}\end{aligned}} \qquad (3.86)$$

The effective mass generated by using the Timoshenko model with the distribution function of Eq. (2.77) is

$$m_{b,e}^{sh} = \cfrac{\begin{aligned}&\rho Rt[\kappa^2 E^2 t^4(0.352R + 0.667w_1)\\&+ \kappa EGR^2 t^2(10.46R + 17.6w_1)\\&+ G^2 R^4(78.941R + 120.69w_1)]\end{aligned}}{(16GR^2 + \kappa Et^2)^2} \qquad (3.87)$$

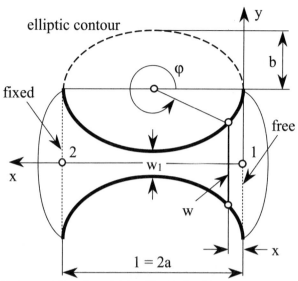

Figure 3.22 Top view of right elliptic microhinge.

The resulting bending-related natural frequency corresponding to a short-beam model is not given here but can be easily determined by means of Eqs. (3.86) and (3.87).

Right elliptic microhinges.   A hinge configuration that is similar to the previous one is the right elliptic microhinge, whose top view is sketched in Fig. 3.22. Approximate compliance equations are presented by Smith et al.[2] whereas exact, closed-form solutions are given by Lobontiu.[4] The elliptic contour is defined by the two semiaxes a and b as well as by the minimum thickness w_1. The following geometric relationships do apply, as seen in Fig. 3.22:

$$x = a + a \cos \varphi \qquad w = w_1 + 2b + 2b \, \sin \varphi \qquad (3.88)$$

where the angle φ is measured as indicated in the same figure and is taken between the limits of π and 2π to cover one semielliptic profile.

The axial stiffness associated with the end that is assumed free, and with respect to the other end that is assumed fixed, is

$$k_{a,e} = \frac{Ebt}{2a[(2b + w_1)/\sqrt{w_1(4b + w_1)} \arctan\sqrt{1 + 4b/w_1} - \pi/4]} \qquad (3.89)$$

The elliptic profile becomes a circular one when $a \to R$ and $b \to R$. By taking these limits into Eq. (3.89), this one changes to Eq. (3.78), which characterizes a right circular hinge, and this proves the validity of Eq. (3.89).

The effective axial mass which is dynamically equivalent to the distributed-parameter inertia of the right elliptic hinge is

$$m_{a,e} = \rho t a(0.352b + 0.667w_1) \qquad (3.90)$$

Again, when $a \to R$ and $b \to R$, and therefore when the elliptic regions become circular, Eq. (3.90) simplifies to Eq. (3.79), which defines the axial effective stiffness of a right circular microhinge. The corresponding axial resonant frequency is

$$\omega_{a,e} = \frac{0.707}{a} \frac{\sqrt{\dfrac{Eb}{0.352b + 0.667w_1}}}{\sqrt{(2b + w_1)\left/\sqrt{w_1(4b + w_1)}\right. \arctan\sqrt{1 + 4b\left/w_1\right.} - \pi\left/4\right.}} \qquad (3.91)$$

The torsional stiffness is found by means of Eq. (3.20) in terms of the axial stiffness. The effective torsion-related mechanical moment of inertia is

$$J_{t,e} = \frac{\rho t a}{12[0.403b^3 + 1.019b^2 w_1 + 0.667w_1(w_1^2 + t^2) + 0.352b(3w_1^2 + t^2)]} \qquad (3.92)$$

Equation (3.92) transforms to Eq. (3.81) for $a \to R$ and $b \to R$, which characterizes a right circular microhinge. The lumped-parameter torsion-related resonant frequency is

$$\omega_{t,e} = \frac{1.414t}{a} \sqrt{\frac{\rho[0.403b^3 + 1.019b^2 w_1 + 0.667w_1(w_1^2 + t^2) + 0.352b(3w_1^2 + t^2)]}{(2b + w_1)\left/\sqrt{w_1(4b + w_1)}\right. \arctan\sqrt{1 + 4b\left/w_1\right.} - \pi/4}} \qquad (3.93)$$

The bending stiffness of a relatively long, right elliptic hinge is

$$k_{b,e} = \frac{Eb^3t^3}{a^3[2.572b^2 + 24.849bw_1 + 4.713w_1^2}$$

$$+ 6(2b + w_1)(4b^2 - 4bw_1 - w_1^2)$$

$$|\sqrt{w_1(4b + w_1)} \arctan\sqrt{1 + 4b/w_1]}$$

(3.94)

When $a \to R$ and $b \to R$, and therefore when the elliptic profile changes into a circular one, Eq. (3.94) transforms indeed to Eq. (3.83), which characterizes a right circular hinge. The effective mass that locates at the microhinge's free end is

$$m_{b,e} = \rho ta(0.308b + 0.471w_1)$$

(3.95)

For $a \to R$ and $b \to R$ the elliptic profile becomes a circular profile, and Eq. (3.95) transforms to Eq. (3.84) which defines a right circular microhinge. The lumped-parameter bending-related resonant frequency is

$$\omega_{b,e} = \frac{bt}{a^2} \frac{\sqrt{\dfrac{Eb}{\rho(0.308b + 0.471w_1)}}}{\sqrt{2.572b^2 + 24.849bw_1 + 4.713w_1^2 + 6(2b + w_1)(4b^2 - 4bw_1 - w_1^2) |\sqrt{w_1(4b + w_1)} \arctan\sqrt{1 + 4b/w_1}}}$$

(3.96)

Short members need to be characterized in terms of both stiffness and effective mass which are calculated by the appropriate short-beam model, as in the recent case with right circular hinge configurations. The resulting equations are quite complex and are not presented here.

Circular corner-filleted microhinges. Another common hinge configuration is the circularly filleted microhinge, whose top view is drawn in Fig. 3.23. It consists of two identical long, circularly filleted microhinges — of the type shown in Fig. 3.17 and dealt with previously — and therefore its lumped-parameter resonant frequencies in axial, torsional, and bending loading can again be derived by means of the two-segment model presented here. This configuration is also utilized in microscale compliant mechanisms as a flexible connector, by mainly employing its in-the-plane bending compliance (bending about the y axis). It derives from the constant rectangular cross-section member, and it has the benefit of including the circular fillet areas which reduce the stress concentration effects and therefore increase the lifetime by

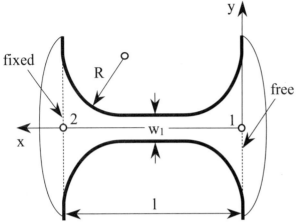

Figure 3.23 Circular corner-filleted microhinge.

improving the structural fatigue response corresponding to oscillatory actuation. Full characterization of this hinge design can be found in Lobontiu[4] and Lobontiu et al.[3] Because in MEMS this design is mainly utilized as a microhinge which bends out of the plane (about the z axis), only that bending motion is studied here, in addition to axial and torsional properties.

The axial stiffness connected to one of the ends (considered free, whereas the other one is considered fixed) is

$$k_{a,e} = \frac{2Etw_1}{2l - 4R - \pi w_1 + 4(2r + w_1) \Big/ \sqrt{1 + 4r/w_1}\, \arctan\sqrt{1 + 4r/w_1}}$$

(3.97)

When $R \to 0$, this equation reduces to Eq. (2.45), which defines the axial stiffness of a constant rectangular cross-section hinge. Also, when $l \to 2R$ and when the configuration of Fig. 3.23 changes to a right circular microhinge such as the one pictured in Fig. 3.21, Eq. (3.97) transforms to Eq. (3.78), as it should. These two limit calculations have been applied to all other lumped-parameter stiffnesses and inertia fractions corresponding to the circular corner-filleted microhinge, and the expected equations characterizing a constant rectangular cross-section or a right circular microhinge have resulted, respectively.

The effective mass associated with free axial vibrations is

$$m_a = \frac{\rho t (5.15 l^2 R^2 - 2.3 l R^3 + 0.88 R^4 + 4 l^3 w_1)}{12 l^2} \tag{3.98}$$

The axial resonant frequency is

$$\omega_{a,e} = 4.9 l \sqrt{\frac{E w_1}{\dfrac{\rho (5.15 l^2 R^2 - 2.3 l R^3 + 0.88 R^4 + 4 l^3 w_1)}{\dfrac{[2 l - 4 R - \pi w_1 + 4 (2 r + w_1)]}{\left[\sqrt{1 + 4 r / w_1}\, \arctan \sqrt{1 + 4 r / w_1}\right]}}}} \tag{3.99}$$

The torsional stiffness is related to the axial stiffness of Eq. (3.97) according to Eq. (3.20). The effective mechanical moment of inertia which is dynamically equivalent to the distributed inertia of the hinge shown in Fig. 3.23 is

$$J_{t,e} = \frac{0.001 \rho t}{l^2} [20 l^3 w_1 (w_1^2 + t^2) - l R^3 (4.57 R^2$$
$$+ 11.5 t^2 + 18.05 r w_1 + 34.51 w_1^2) + R^4 (0.83 R^2$$
$$+ 4.38 t^2 + 4.57 R w_1 + 13.14 w_1^2) + l^2 R^2 (26.28 R^2$$
$$+ 25.75 t^2 + 69.03 R w_1 + 77.26 w_1^2)] \tag{3.100}$$

The torsional resonant frequency is

$$\omega_{t,e} = 21.91 l t \sqrt{\frac{\dfrac{G w_1}{\dfrac{\rho [2 l - 4 R - \pi w_1 + 4 (2 r + w_1)]}{\left[\sqrt{1 + 4 r / w_1}\, \arctan \sqrt{1 + 4 r / w_1}\right]}}}{20 l^3 w_1 (w_1^2 + t^2) - l R^3 (4.57 R^2 + 11.5 t^2}} } \\ {} $$

$$\begin{aligned} &+ 18.05 r w_1 + 34.51 w_1^2) + R^4 (0.83 R^2 \\ &+ 4.38 t^2 + 4.57 R w_1 + 13.14 w_1^2) \\ &+ l^2 R^2 (26.28 R^2 + 25.75 t^2 \\ &+ 69.03 R w_1 + 77.26 w_1^2) \end{aligned} \tag{3.101}$$

In bending, the lumped-parameter stiffness corresponding to a long configuration is

$$k_{b,e} = \frac{2Et^3}{8(l-2R)(l^2-lR+R^2)/w_1 - 6(3.14l^2}$$

$$-2.28lR - 8.86R^2) + 49.7Rw_1 + 9.42w_1^2$$

$$+12(l-2R)(2R+w_1)\ln(1+2R/w_1) \tag{3.102}$$

$$-12(2R+w_1)(w_1^2 + 4w_1 R - 4R^2$$

$$+4lR - 2l^2)/\sqrt{w_1(4R+w_1)}\arctan\sqrt{1+4R/w_1}$$

The effective bending mass is

$$m_{b,e} = \rho t \left[\frac{\begin{array}{c} 0.002R^2(209.92l^6 - 140.67l^5 R \\ +40.17l^4 R^2 + 8.7l^3 R^3 + 3.67l^2 R^4 \\ -4.53lR^5 + R^6) \end{array}}{l^6} + 0.24lw_1 \right] \tag{3.103}$$

The bending-related resonant frequency is too complex to be explicitly given, but it can readily be found by means of Eqs. (3.102) and (3.103).

3.3.4 Circularly notched microcantilevers

The circularly notched microcantilever design, which was first presented by Garcia et al.,[19] is formed of a right circular portion connected in series to a constant-cross-section segment, as shown in Fig. 3.24.

The axial stiffness is

$$k_{a,e} = \frac{4Etw_1}{4l_1 - \pi w_1 + 4w_1(2R+w_2)}$$

$$\frac{}{/\sqrt{w_2(4R+w_2)}\arctan\sqrt{1+4R/w_2}} \tag{3.104}$$

Equation (3.104) reduces to Eq. (2.45) when $R \to 0$, which shows that the compound configuration reduces to a constant rectangular cross-section microcantilever when the circularly notched segment vanishes. The effective axial inertia is

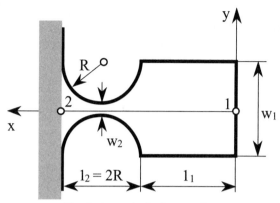

Figure 3.24   Circularly notched microcantilever.

$$m_{a,e} = \frac{\rho t(16.88R^4 + 4l_1^3 w_1 + 24l_1^2 R w_1 + 48l_1 R^2 w_1 + 32R^3 w_2)}{12(l_1 + 2R)^2} \tag{3.105}$$

This equation, too, reduces to that corresponding to a constant-cross-section cantilever of length l_1 [Eq. (2.49)] for $R \to 0$. The axial resonant frequency is

$$\omega_{a,e} = 6.93(l_1 + 2R)\sqrt{\frac{Ew_1}{\frac{\rho\left[4l_1 - \pi w_1 + 4w_1(2R + w_2) \middle/ \sqrt{w_2(4R + w_2)} \arctan\sqrt{1 + 4R/w_2}\right]}{16.88R^4 + 4l_1^3 w_1 + 24l_1^2 R w_1 + 48l_1 R^2 w_1 + 32R^3 w_2}}} \tag{3.106}$$

The torsional stiffness is related to the axial stiffness, according to Eq. (3.20). The effective mechanical moment of inertia is calculated as

$$J_{t,e} = \frac{\rho t}{12}\Big\{0.44R^4 + l_1 w_1(l_1^2 + 6l_1 R + 12R^2)(w_1^2 + t^2) \\ \Big/ [3(l_1 + 2R)^2] + 1.15R^3 w_2 + R w_2(w_2^2 + t^2) \\ + 0.43R^2(t^2 + 3w_2^2)\Big\} \tag{3.107}$$

When $R \to 0$, Eq. (3.107) simplifies to Eq. (2.55), which gives the effective mechanical moment of inertia for a constant rectangular cross-section cantilever. The lumped-parameter torsional resonant frequency is

$$
\omega_{t,e} = 4t \sqrt{\dfrac{\dfrac{Gw_1}{\rho\left[4l_1 - \pi w_1 + 4w_1(2R + w_2)\right.}{\left/\sqrt{w_2(4R + w_2)}\,\arctan\sqrt{1 + 4R/w_2}\,\right]}}{\sqrt{\begin{array}{l}0.44R^4 + l_1 w_1(l_1^2 + 6l_1 R + 12R^2)(w_1^2 + t^2) \\ \left/[3(l_1 + 2R)^2] + 1.15R^3 w_2 + Rw_2(w_2^2 + t^2)\right. \\ +0.43R^2(t^2 + 3w_2^2)\end{array}}}}
\tag{3.108}
$$

For long configurations, the bending stiffness is

$$
\begin{aligned}
k_{b,e} = \; & 2Et^3 w_1 \Big/ \Big[8l_1^3 - 37.7l_1 w_1(l_1 + 2R) + 5.15w_1 R^2 \\
& + 49.7Rw_1 w_2 + 9.42w_1 w_2^2 \\
& + 12w_1(2R + w_2)\big(l_1(l_1 + 2R)\{\,2\,\arctan\big[2R \\
& \left/\sqrt{w_2(4R + w_2)}\,\right] + \pi\} + (4R^2 - 4Rw_2 - w_2^2) \\
& \arctan\sqrt{1 + 4R/w_2}\,\big)\Big/\sqrt{w_2(4R + w_2)}
\end{aligned}
\tag{3.109}
$$

When $R \to 0$, Eq. (3.109) simplifies to Eq. (2.61), which corresponds to a constant-cross-section cantilever of length l_1. The effective mass that is located at the free end of the microcantilever and is dynamically equivalent to the distributed-parameter inertia of the bending vibrating member is

$$
\begin{aligned}
m_{b,e} = \; & \rho t[l_1^3 w_1(0.236l_1^4 + 3.3l_1^3 R + 19.8l_1^2 R^2 + 66l_1 R^3 \\
& + 132R^4) + l_1^2 R^5(10.22R + 144w_1 \\
& + 13.95w_2) + R^7(19.66R + 28.84w_2) \\
& + l_1 R^6(28.34R + 64w_1 + 40.05w_2)] \Big/ (l_1 + 2R)^6
\end{aligned}
\tag{3.110}
$$

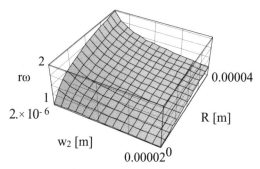

Figure 3.25 Constant cross section to circularly notched comparison in terms of the bending resonant frequency.

Again, when $R \to 0$, Eq. (3.110) reduces to Eq. (2.66), which provides the effective inertia of a constant rectangular cross-section cantilever of length l_1. The bending resonant frequency can be expressed by using the lumped-parameter stiffness [Eq. (3.109)] and effective mass [Eq. (3.110)] but is too complex and is not included here.

Example: Compare the bending resonant frequency of a constant rectangular cross-section microcantilever to that of a circularly notched microcantilever such as the one in Fig. 3.24. Assume the two designs have the same length $l = 1000$ μm, width $w_1 = 100$ μm, and thickness t and are built of the same material.

The bending resonant frequency of a constant rectangular cross-section microcantilever is given in Eq. (2.68), whereas the similar frequency for the configuration illustrated in Fig. 3.24 can be found by means of the stiffness [Eq. (3.109)] and effective mass [Eq. (3.110)], and Fig. 3.25 is the three-dimensional plot of the resonant frequency ratio (constant cross-section to circularly notched). The bending frequency ratio (denoted by $r\omega$) increases for relatively thin notches (smaller values of w_2) and large fillet radii, as indicated in Fig. 3.25 (where $r\omega$ is the bending frequency ratio).

3.4   Hollow Microcantilevers

Microcantilevers in hollow configurations are generally utilized as substitutes for their solid counterparts in cases where reduction of the main bending stiffness and subsequent decrease of the associated resonant frequency are needed. Two hollow designs, which have also been presented by Lobontiu and Garcia,[5] are studied here, namely, the rectangular and the trapezoid microcantilevers.

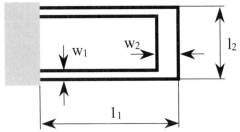

Figure 3.26   Top view of hollow rectangular microcantilever.

3.4.1 Rectangular microcantilevers

Figure 3.26 shows the top view of a hollow microcantilever of rectangular envelope. It consists of two arms that are flexible and a connecting arm which can be considered rigid.

Under the assumptions that only the thinner side bars are compliant, the resulting stiffness of the structure, be it axial, torsional, or bending-related, will be twice the corresponding stiffness of a single side bar, because the two flexible parts behave as two springs in parallel. The axial stiffness of the hollow rectangular microcantilever is thus

$$k_{a,e} = \frac{2Ew_1 t}{l_1} \tag{3.111}$$

The lumped-parameter mass which corresponds to the distributed parameter inertia of the axial free vibrations consists of three fractions: two identical terms that result from the flexible parts and the mass of the rigid transverse member. Such situations, where effective mass factions coming from flexible parts are added to actual masses of rigid members, is studied more thoroughly in subsequent chapters of this book. The total mass undergoing free axial vibrations is

$$m_{a,e} = \frac{2}{3}m_{1,a} + m_2 = \rho t \left(\frac{2l_1 w_1}{3} + l_2 w_2 \right) \tag{3.112}$$

The axial-related resonant frequency is

$$\omega_{a,e} = 2.45 \sqrt{\frac{Ew_1}{\rho l_1 (2l_1 w_1 + l_2 w_2)}} \tag{3.113}$$

The torsional stiffness of this hollow configuration (its compliance is given by Lobontiu and Garcia[5]) is

$$k_{t,e} = \frac{(4Gl_1^2 + 3El_2^2)w_1 t^3}{6l_1^3}$$ (3.114)

The effective mechanical moment of inertia, which is dynamically equivalent to the distributed inertia of the hollow rectangular microcantilever undergoing free torsional vibrations, is

$$J_{t,e} = \frac{2J_{t1}}{3} + J_{t2} = \frac{\rho t[2l_1 w_1(w_1^2 + t^2) + 3l_2 w_2(w_2^2 + t^2)]}{36}$$ (3.115)

The torsion resonant frequency is

$$\omega_{t,e} = \frac{2.45t}{l_1} \sqrt{\frac{(4Gl_1^2 + 3El_2^2)w_1}{\rho l_1[2l_1 w_1(w_1^2 + t^2) + 3l_2 w_2(w_2^2 + t^2)]}}$$ (3.116)

The bending stiffness is

$$k_{b,e} = \frac{ew_1 t^3}{2l_1^3}$$ (3.117)

The effective mass in bending is

$$m_{b,e} = 2\frac{33}{140}m_{b1} + m_2 = \rho t\left(\frac{33l_1 w_1}{70} + l_2 w_2\right)$$ (3.118)

and the bending-related resonant frequency is

$$\omega_{b,e} = \frac{5.92t}{l_1} \sqrt{\frac{Ew_1}{\rho l_1(33l_1 w_1 + 70l_2 w_2)}}$$ (3.119)

Example: Perform a comparison similar to that of the previous example, by studying the bending resonant frequency of a solid constant rectangular cross-section microcantilever in contrast to that of a hollow configuration. It is assumed that both configurations have the same geometric envelope, and identical thicknesses, and that they are built from the same material.

The sought frequency ratio is determined by using Eqs. (2.67) and (3.119), which give the bending resonant frequencies of a solid constant rectangular cross-section microcantilever and of a hollow rectangular one, respectively. By using the nondimensional parameters:

$$c_1 = \frac{w_1}{l_2} \qquad c_2 = \frac{w_2}{l_1}$$ (3.120)

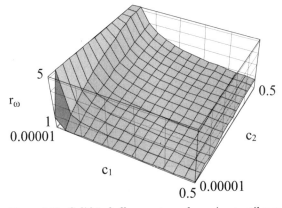

Figure 3.27 Solid-to-hollow rectangular microcantilever comparison in terms of the bending resonant frequency.

the solid-to-hollow bending resonant frequency ratio can be expressed as

$$r_\omega = \frac{\omega_s}{\omega_h} = 0.174 \sqrt{\frac{33c_1 + 70c_2}{c_1}} \qquad (3.121)$$

and Fig. 3.27 is the three-dimensional plot of this frequency ratio.

The figure reveals the more substantive impact that parameter c_1 (which connects the in-plane thickness w_1 to l_2) has on the frequency ratio. For small values of this parameter, the corresponding curves in Fig. 3.27 are strongly nonlinear, and small increases in c_1 result in large decreases in the frequency ratio. The influence of parameter c_2 is more discrete as it measures contributions to the overall mass in the hollow configuration, and not so much the interventions in the bending stiffness.

3.4.2   Trapezoid microcantilevers

The top view with the defining geometric parameters of a hollow trapezoid microcantilever is sketched in Fig. 3.28. The assumption is also made that the side arms are flexible, whereas the transverse connecting arm is rigid. Compliances in axial loading, torsion, and bending are provided for this configuration by Lobontiu and Garcia.[5]

In calculating the significant resonant frequencies, the inertia fractions corresponding to axial, torsion, and bending vibrations that have been determined for hollow rectangular microcantilevers are also valid for trapezoid configurations.

The axial stiffness is

$$k_{a,e} = \frac{2Ew_1 t}{l_1 \cos \alpha} \qquad (3.122)$$

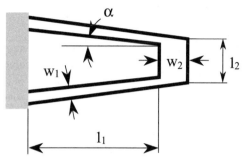

Figure 3.28   Top view of a hollow trapezoid microcantilever.

The axial resonant frequency is

$$\omega_{a,e} = 2.45\sqrt{\frac{Ew_1}{\rho l_1 (2l_1 w_1 + 3l_2 w_2)\cos\alpha}}$$ (3.123)

The torsional stiffness is

$$k_{t,e} = \frac{2EGw_1 t^3}{3l_1(E\cos^2\alpha + 4G\sin^2\alpha)}$$ (3.124)

The torsional resonant frequency is

$$\omega_{t,e} = 4.9t\sqrt{\frac{EGw_1}{\rho l_1[2l_1 w_1(w_1^2 + t^2) + 3l_2 w_2(w_2^2 + t^2)](E\cos^2\alpha + 4G\sin^2\alpha)}}$$ (3.125)

The bending stiffness is

$$k_{b,e} = \frac{8EGw_1 t^3}{l_1\{16Gl_1^2 + 3l_2[El_2\cos^2\alpha + 4G(2l_1 + l_2\sin\alpha)\sin\alpha]\}}$$ (3.126)

The bending resonant frequency is

$$\omega_{b,e} = 2.83t\sqrt{\frac{EGtw_1}{\rho t l_1(33l_1 w_1 / 70 + l_2 w_2)\{16Gl_1^2 + 3l_2[El_2\cos^2\alpha + 4G(2l_1 + l_2\sin\alpha)\sin\alpha]\}}}$$ (3.127)

3.5 Sandwiched Microcantilevers (Multimorphs)

Sandwiched components such as microcantilevers or microbridges are often utilized in microtransduction (actuation and/or sensing) where one layer achieves the structural and elastic recovery functions (most often the silicon or polysilicon) and the other layers are active in the sense that they deform under activation or environmental stimuli. The resonant frequencies corresponding to axial deformation, torsion, and loading are presented next for equal-length microcomponents and for dissimilar-length ones.

3.5.1 Microcantilevers of equal-length layers

In equal-length multimorphs, the component layers have identical lengths. Figure 3.29 shows a microcantilever consisting of n layers. The resonant frequencies are determined next for this micromember. The resonant frequencies for a multimporph are very similar to those of a homogeneous micromember having the same length.

In the case of axial vibrations, for instance, an equivalent rigidity needs to be used in the form (see Lobontiu and Garcia[5]):

$$k_{a,e} = \sum_{i=1}^{n} \frac{E_i A_i}{l} \tag{3.128}$$

where l is the identical length and A_i is the cross-sectional area of the ith component. In a similar manner, it can be shown that the lumped-parameter mass which needs to be placed at the microcantilever's free tip, in order to generate the same inertial effect as the distributed mass of the multimorph, according to Rayleigh's principle, is

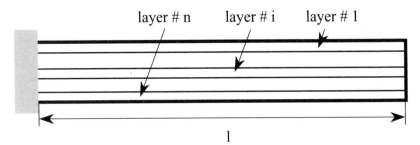

Figure 3.29 Sandwiched microcantilever.

$$m_{a,e} = l \sum_{i=1}^{n} \frac{\rho_i A_i}{3} = \frac{m}{3} \tag{3.129}$$

where ρ_i is the material density of the ith component. As a consequence, the axial resonant frequency of a multimorph can be expressed as

$$\omega_{a,e} = \sqrt{\frac{k_{a,e}}{m_{a,e}}} = \frac{\sqrt{3}}{l} \sqrt{\frac{\sum_{i=1}^{n} E_i A_i}{\sum_{i=1}^{n} \rho_i A_i}} \tag{3.130}$$

In the case of a two-component sandwiched microcantilever (bimorph), the axial resonant frequency becomes

$$\omega_{a,e} = \frac{\sqrt{3}}{l} \sqrt{\frac{E_1 t_1 + E_2 t_2}{\rho_1 t_1 + \rho_2 t_2}} \tag{3.131}$$

By considering that

$$t_2 = c t_1 \tag{3.132}$$

Eq. (3.131) changes to

$$\omega_{a,e} = \frac{\sqrt{3}}{l} \sqrt{\frac{E_1 + c E_2}{\rho_1 + c \rho_2}} \tag{3.133}$$

When $c \to 0$ (corresponding to the situation where the microcantilever is homogeneous and made up of a single layer), Eq. (3.133) reduces to

$$\omega_a = \sqrt{3} \sqrt{\frac{E_1 A_1}{m_1 l}} \tag{3.134}$$

which is the known equation for the axial resonant frequency of a homogeneous microcantilever.

A similar reasoning is now applied to the free torsional vibrations of a multimorph whose equivalent rigidity is

$$k_{t,e} = \sum_{i=1}^{n} \frac{G_i I_{ti}}{l} \tag{3.135}$$

where G_i is the shear modulus and I_{ti} is the torsional moment of inertia for the ith component and was defined previously for a homogeneous, constant rectangular cross-section microcantilever. For very thin components, as the case is with the majority of MEMS, the torsional moment of inertia is

$$I_{ti} = \frac{wt_i^3}{3} \qquad (3.136)$$

It has also been shown (see Lobontiu,[4] for instance) that the equivalent mechanical moment of inertia $J_{t,e}$ is calculated as

$$J_{t,e} = \frac{J_t}{3} \qquad (3.137)$$

where J_t is the mechanical moment of inertia of the whole multimorph and can be calculated as

$$J_t = \sum_{i=1}^{n} [J_{ti} + m_i(z_i - z_C)^2] \qquad (3.138)$$

The mechanical moment of inertia of the ith component taken with respect to its central axis is

$$J_{ti} = \frac{lwt_i\rho_i(w^2 + t_i^2)}{12} \qquad (3.139)$$

The distance positioning the central axis of the ith component measured from the multimorph lowest layer (layer n), as suggested in Fig. 3.30, is

$$z_i = \frac{t_i}{2} + \sum_{k=i+1}^{n} t_k \qquad (3.140)$$

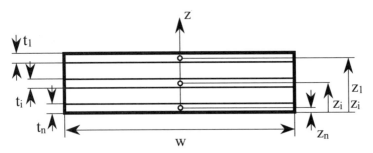

Figure 3.30 Cross section of sandwich bar with main geometry parameters for torsional resonant frequency calculations.

whereas the distance corresponding to the geometric symmetry center of the compound cross section is

$$z_C = \frac{\sum\limits_{i=1}^{n} z_i}{n} \qquad (3.141)$$

Example: Establish the errors generated in calculating the torsional mechanical moment of inertia by the simplified equation

$$J_t' = \sum_{i=1}^{n} J_{ti} \qquad (3.142)$$

instead of calculating it by means of Eq. (3.138) for a multimorph formed of three layers. Consider that $\rho_2 = 1.1\rho_1$, $\rho_3 = 0.9\rho_1$, $w = 50$ µm, and $t_1 = 3$ µm. The following nondimensional amounts are utilized:

$$c_2 = \frac{t_2}{t_1} \qquad c_3 = \frac{t_3}{t_1} \qquad (3.143)$$

which enable is to express the error

$$e J_t = \frac{J_t - J_t'}{J_t'} \qquad (3.144)$$

just in terms of these two parameters, as plotted in Fig. 3.31. As this figure indicates, the relative errors between the two expressions of the torsional moment of inertia are less than 4 percent when the thickness parameters of layers 2 and 3, c_2 and c_3, span the range $0.5 \to 1.5$. For very thin layers, this

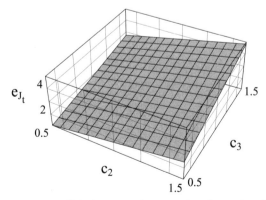

Figure 3.31 Relative errors in expressing the torsional mechanical moment of inertia.

thickness range is quite extended, and therefore errors smaller than the maximum value shown on the plot of Fig. 3.31 can be expected.

The torsional resonant frequency can therefore be written as

$$\omega_{t,e} = \sqrt{\frac{k_{t,e}}{m_{t,e}}} = \sqrt{3}\sqrt{\sum_{i=1}^{n}\frac{G_i I_{ti}}{J_t l}}$$
(3.145)

For a bimorph (two-component sandwiched microcantilever), the torsional resonant frequency becomes

$$\omega_{t,e} = \frac{2\sqrt{3}}{l}\sqrt{\frac{G_1 t_1^3 + G_2 t_2^3}{\rho_1 t_1(t_1^2 + 3t_2^2 + w^2) + \rho_2 t_2(t_2^2 + 3t_1^2 + w^2)}}$$
(3.146)

When $t_2 \to 0$, Eq. (3.146) simplifies to

$$\omega_t = \frac{2\sqrt{3}t_1}{l}\sqrt{\frac{G_1}{\rho_1(t_1^2 + w^2)}} = \sqrt{\frac{3k_{t1}}{J_{t1}}}$$
(3.147)

which is the known torsion-related resonant frequency of a homogeneous (one-component) microbar.

The first bending resonant frequency is now calculated based on a lumped-parameter model and on Figs. 3.30 and 3.32. The long-beam (Euler-Bernoulli) model is first approached. As shown by Lobontiu and Garcia,[5] the equivalent bending rigidity is calculated as

$$(EI_y)_e = \sum_{i=1}^{n} E_i\left[I_{yi} + z_i A_i(z_i - z_N)\right]$$
(3.148)

where I_{yi} is the moment of inertia of the ith component with respect to its central axis y_i. The position of the neutral axis z_N is calculated as

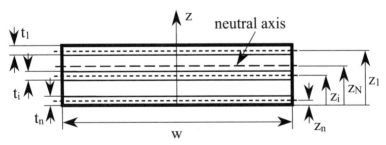

Figure 3.32 Cross section of sandwich beam with main geometry parameters for the bending resonant frequency calculation.

$$z_N = \frac{\sum\limits_{i=1}^{n} z_i E_i A_i}{\sum\limits_{i=1}^{n} E_i A_i} \tag{3.149}$$

The stiffness of this equivalent cantilever beam is

$$k_{b,e} = \frac{3(EI_y)_e}{l^3} \tag{3.150}$$

It can also be shown that the mass fraction that has to be placed at the free end of the microcantilever and that is dynamically equivalent to the distributed mass of the multimorph is

$$m_{b,e} = \frac{33l \sum\limits_{i=1}^{n} (\rho_i A_i)}{140} = \frac{33}{140} m \tag{3.151}$$

where m is the mass of the whole multimorph. The first bending resonant frequency is calculated by means of Eqs. (3.150) and (3.151) as

$$\omega_{b,e} = \sqrt{\frac{k_{b,e}}{m_{b,e}}} = 3.567 \sqrt{\frac{(EI_y)_e}{ml^3}} \tag{3.152}$$

In the case of a bimorph, Eq. (3.152) reduces to

$$\omega_{b,e} = \frac{1.03\sqrt{E_1^2 t_1^4 + E_2^2 t_2^4 + 2E_1 E_2 t_1 t_2 (2t_1^2 + 3t_1 t_2 + 2t_2^2)}}{\sqrt{(E_1 t_1 + E_2 t_2)(\rho_1 t_1 + \rho_2 t_2)}/l^2} \tag{3.153}$$

When we express t_2 in terms of t_1 according to Eq. (3.132), then Eq. (3.153) simplifies to

$$\omega_{b,e} = \frac{1.03 t_1}{l^2} \sqrt{\frac{E_1^2 + 2c[2 + c(3 + 2c)]E_1 E_2 + c^4 E_2^2}{(E_1 + cE_2)(\rho_1 + c\rho_2)}} \tag{3.154}$$

When $c \to 0$ (the bimorph becomes a unimorph formed of only material 1), Eq. (3.154) simplifies to

$$\omega_b = 3.567 \sqrt{\frac{E_1 I_{y1}}{m_1 l^3}} \tag{3.155}$$

which is the known relationship.

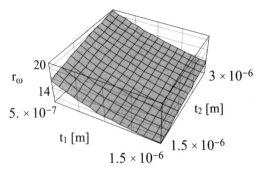

Figure 3.33 Torsional-to-bending resonant frequency ratio in terms of layer thicknesses.

Example: Study the influence of geometry on the bending and torsional resonant frequency of a bimorph when the following parameters are known: $E_1 = 150$ GPa, $G_1 = 60$ GPa, $E_2 = 180$ GPa, $G_2 = 72$ GPa. The densities of the two materials are equal. Consider that the geometric parameters range as follows: $t_1 \to [0.5\ \mu m, 1.5\ \mu m]$, $t_2 \to [1.5\ \mu m, 3\ \mu m]$, $l \to [800\ \mu m, 1500\ \mu m]$, $w \to [50\ \mu m, 150\ \mu m]$.

The torsion-to-bending resonant frequency ratio is formed by using Eqs. (3.146) and (3.153). By first considering that $l = 1200\ \mu m$ and $w = 100$ μm, the thickness parameters are allowed to vary within their specified ranges, and the plot of the resonant frequency ratio r_ω in Fig. 3.33 is obtained.

For the thickness ranges and all other constant material and geometry parameter values, the torsional resonant frequency is 14 to 20 times higher than the bending resonant frequency, as Fig. 3.33 indicates. The plot also shows that the torsion-to-stiffness resonant frequency ratio decreases with the increase of the thickness of the deposited (thinner) layer 1 and increases with the increase of the thickness of the substrate (thicker) layer.

A similar numerical simulation is presented in Fig. 3.34, where the following values have been selected for the layer thicknesses: $t_1 = 1\ \mu m$ and $t_2 = 2\ \mu m$, whereas the parameters l and w have been allowed to vary within their specified ranges.

Varying the length and width of the sandwich makes the torsion-to-bending resonant frequency ratio vary from 10 to 30, as shown in Fig. 3.34. Increasing the length and decreasing the width of the bimorph contribute to increasing the resonant frequency ratio.

For relatively short microcantilevers, Timoshenko's model needs to be employed to determine the first bending-related resonant frequency. The stiffness of a multimorph can be put in a form similar to that given in Eq. (2.79) for a homogeneous microcantilever, namely,

$$k_{b,e} = \frac{3(EI_y)_e (GA)_e}{l[(GA)_e l^2 + 3\kappa(EI_y)_e]} \tag{3.156}$$

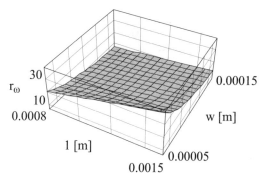

Figure 3.34 Torsional-to-bending resonant frequency ratio in terms of layer length and width.

The equivalent bending rigidity $(EI_y)_e$ is expressed in Eq. (3.148). The shear-related rigidity $(GA)_e$ can be found in a similar manner to the axial and torsional rigidities, namely,

$$(GA)_e = \sum_{i=1}^{n} G_i A_i \qquad (3.157)$$

The equivalent mass is similar to the one expressed in Eq. (2.80) for a homogeneous microcantilever, namely,

$$m_{b,e} = \frac{3m[140\kappa^2(EI_y)_e^2 + 77\kappa(EI_y)_e(GA)_e l^2 + 11(GA)_e^2 l^4]}{140[(GA)_e l^2 + 3\kappa(EI_y)_e]^2} \qquad (3.158)$$

The first bending-related resonant frequency of a short multimorph is therefore

$$\omega_{b,e} = \frac{11.832\sqrt{(EI_y)_e(GA)_e[(GA)_e l^2 + 3\kappa(EI_y)_e]\big/(l\,m)}}{\sqrt{140\kappa^2(EI_y)_e^2 + 77\kappa(EI_y)_e(GA)_e l^2 + 11(GA)_e^2 l^4}} \qquad (3.159)$$

The resonant frequency of a bimorph can be obtained from Eq. (3.159) by employing only two components in the bending and shearing rigidities, but the final equation is too complicated and is not provided here.

3.5.2 Microcantilevers of dissimilar-length layers

Bimorphs can be microfabricated for use in transduction by attaching a patch over a substrate such that the lengths of the two microcomponents are not equal, although their widths w are equal. Figure 3.35 is

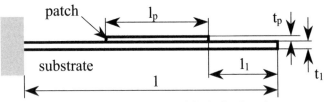

Figure 3.35 Bimorph microcantilever of dissimilar-length components.

the side view of the bimorph microcantilever with dissimilar-length components. We calculate the natural frequencies corresponding to axial, torsional, and bending vibrations by lumping the stiffness and mass (inertia) at the bimorph's free tip, as shown in the following.

In axial free vibrations, the stiffness of the patched microcantilever can be calculated by considering that three portions are connected in series, namely, the tip portion of length l_1, the composite portion (substrate cantilever and patch) of length l_p, and the root portion. By applying the series connection rule of springs

$$\frac{1}{k_{a,e}} = \frac{l_1}{E_1 A_1} + \frac{l_p}{E_1 A_1 + E_p A_p} + \frac{l - (l_1 + l_p)}{E_1 A_1} \tag{3.160}$$

the axial stiffness can be expressed as

$$k_{a,e} = \frac{E_1 t_1 (E_1 t_1 + E_p t_p) w}{E_1 t_1 l + E_p t_p (l - l_p)} \tag{3.161}$$

Equation (3.161) reduces to

$$k_a = \frac{E_1 t_1 w}{l} \tag{3.162}$$

when $t_p \to 0$, which is the known relationship for a one-component microrod.

The lumped mass which corresponds to axial vibrations is calculated by means of Rayleigh's principle, a procedure that has been discussed in detail previously. Its equation is

$$m_{a,e} = \frac{m_1 + m_p [3(l - l_1)^2 - 3(l - l_1) l_p + l_p^2] / l^2}{3} \tag{3.163}$$

For $m_p \to 0$ (no patch on the microcantilever), Eq. (3.163) reduces to

$$m_a = \frac{m_1}{3} \tag{3.164}$$

which is the known relationship for a one-component microrod. By combining Eqs. (3.161) and (3.163), the resonant frequency corresponding to free axial vibrations of the patched microcantilever becomes

$$\omega_{a,e} = \sqrt{\frac{k_{a,e}}{m_{a,e}}}$$

$$= \frac{1.73l\sqrt{E_1 t_1 (E_1 t_1 + E_p t_p) / [E_1 t_1 l + E_p t_p (l - l_p)]}}{\sqrt{l^3 t_1 \rho_1 + l_p [3(l - l_1)^2 - 3(l - l_1)l_p + l_p^2] t_p \rho_p}} \qquad (3.165)$$

In torsion, similar to the axial problem, the stiffness of the patched microcantilever can be found by considering the series connection between the three portions above-mentioned. The stiffness equation is

$$k_{t,e} = \frac{G_1 t_1^3 (G_1 t_1^3 + G_p t_p^3) w}{3[G_1 l t_1^3 + G_p (l - l_p) t_p^3]} \qquad (3.166)$$

For $l_p \to 0$ and $t_p \to 0$, Eq. (3.166) reduces to

$$k_t = \frac{G_1 w t_1^3}{3l} \qquad (3.167)$$

which is the stiffness of a homogeneous fixed-free bar.

The equivalent torsional mechanical moment of inertia is determined again by means of Rayleigh's principle, and its equation is

$$J_{t,e} = \frac{w\left\{l^3 \rho_1 t_1 (t_1^2 + w^2) + 3l^2 l_p t_p [3\rho_1 t_1 t_p + \rho_p (3t_1^2 + t_p^2 + w^2)] \right.}{36l^2}$$

$$\left. -3l l_p t_p (2l_1 + l_p)[3\rho_1 t_1 t_p + \rho_p (3t_1^2 + t_p^2 + w^2)] \right.$$

$$\left. +l_p t_p (3l_1^2 + 3l_1 l_p + l_p^2)[3\rho_1 t_1 t_p + \rho_p (3t_1^2 + t_p^2 + w^2)]\right\}}{} \qquad (3.168)$$

When $l_p \to 0$ and $t_p \to 0$, Eq. (3.168) simplifies to

$$J_t = \frac{1}{3} \frac{\rho_1 l w t_1 (w^2 + t_1^2)}{12} \qquad (3.169)$$

which is the equation expressing the equivalent mechanical moment of inertia in torsion for a homogeneous fixed-free bar.

The torsional resonant frequency of the bar is calculated by Eqs. (3.166) and (3.168), and its equation is

$$\omega_{t,e} = 6lt_1\sqrt{\frac{G_1t_1(G_1t_1^3 + G_pt_p^3)}{c_t}}$$ (3.170)

where

$$c_t = 3[G_1lt_1^3 + G_p(l - l_p)t_p^3]\{\rho_1l^3t_1(t_1^2 + w^2)$$

$$+ 3l^2l_pt_p[3\rho_1t_1t_p + \rho_p(3l_1^2 + t_p^2 + w^2)]$$

$$- 3ll_pt_p(2l_1 + l_p)[3\rho_1t_1t_p + \rho_p(3t_1^2 + t_p^2 + w^2)]$$

$$+ l_pt_p(3l_1^2 + 3l_1l_p + l_p^2)[3\rho_1t_1t_p + \rho_p(3t_1^2 + t_p^2 + w^2)]\}$$ (3.171)

The lumped-parameter stiffness in bending is determined by calculating the tip displacement of the sandwiched microcantilever which is produced by a tip force, by means of Castigliano's displacement theorem, for instance, as shown in Lobontiu and Garcia.[5] This stiffness is

$$k_{b,e} = \frac{w}{4\{[l^3 - l_p(3l_1^2 + 3l_1l_p + l_p^2)]/(E_1t_1^3)}$$

$$+[l_p(3l_1^2 + 3l_1l_p + l_p^2)(E_1t_1 + E_pt_p)]$$ (3.172)

$$/[E_1^2t_1^4 + E_p^2t_p^4 + 2E_1E_pt_1t_p(2t_1^2 + 3t_1t_p + 2t_p^2)]\}$$

Equation (3.172) simplifies to the known relationship

$$k_b = \frac{E_1wt_1^3}{4l^3} = \frac{3EI_y}{l^3}$$ (3.173)

when $l_p \to 0$ and $t_p \to 0$, a relationship that corresponds to a homogeneous (one-component) microcantilever.

In bending, the lumped-parameter mass which is placed at the free end of the microcantilever is calculated by Rayleigh's principle in a way that has been described previously here. By equating the kinetic energy of the distributed-parameter bimorph to that of an equivalent mass placed at the free end of a massless cantilever, the inertia fraction is

$$m_{b,e} = \frac{33(m_1 + l_p'/l_pm_p)}{140}$$ (3.174)

where

$$l_p' = \left[140 l_p + \sum_{\substack{i=1 \\ i \neq 5}}^{6} c_i \frac{(l_1 + l_p)^{i+1} - l_1^{i+1}}{l^i} \right] \Bigg| 33 \tag{3.175}$$

with $c_1 = -210$, $c_2 = 105$, $c_3 = 35$, $c_4 = -42$, and $c_6 = 5$. Equation (3.174) reduces to

$$m_b = \frac{33 m_1}{140} \tag{3.176}$$

in the case where there is no patch on the microcantilever ($m_p = 0$), and this is the known relationship, giving the equivalent mass of a homogeneous microcantilever.

The bending-related resonant frequency is now computed by means of Eqs. (3.172) and (3.174) as

$$\omega_{b,e} = 1.03 \sqrt{\frac{w l_p}{(l_p m_1 + l_p' m_p) c_b}} \tag{3.177}$$

with

$$c_b = \frac{l^3 - l_p (3 l_1^2 + 3 l_1 l_p + l_p^2)}{E_1 t_1^3}$$

$$+ \frac{l_p (3 l_1^2 + 3 l_1 l_p + l_p^2)(E_1 t_1 + E_p t_p)}{E_1^2 t_1^4 + E_p^2 t_p^4 + 2 E_1 E_p t_1 t_p (2 t_1^2 + 3 t_1 t_p + 2 t_p^2)} \tag{3.178}$$

Obviously, Eq. (3.177) reduces to that equation giving the bending resonant frequency of a homogeneous microcantilever when $l_p \to 0$, $t_p \to 0$, and $m_p \to 0$.

Note: Multimorph microcomponents, in either the equal- or dissimilar-length configurations, are generally constructed of thin layers (the substrate included), and therefore shearing effects (and the corresponding model alterations imposed by the Timoshenko model) can be ignored.

Example: Study how the patch position and length (quantified by l_1 and l_p in Fig. 3.35) of a bimorph microcantilever influence the bending resonant frequency. Given: $E_2 = 150$ GPa, $E_1 = 120$ GPa, $\rho_1 = 2300$ kg/m^3, $\rho_2 = 2500$ kg/m^3, $l = 1500$ µm, and $w = 100$ µm. Consider that the monitored parameters range within the following intervals: $l_1 \to [800 \text{ µm}, 2000 \text{ µm}]$, $l_p \to [50 \text{ µm}, 100 \text{ µm}]$.

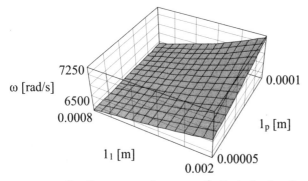

Figure 3.36 Bending resonant frequency of a dissimilar-length sandwich microcantilever.

The bending resonant frequency of Eq. (3.177) is plotted in Fig. 3.36 for the parameters given in this example. Figure 3.36 indicates that the bending resonant frequency of a bimorph increases with a larger patch length, and this is due to the fact that the patch stiffness increases more compared to the patch mass with length (third power and first power, respectively). Also, Fig. 3.36 suggests that for a constant patch length, the resonant frequency is higher for patch positions that are closer to the microcantilever root (which means larger l_1), and this also makes physical sense since the effective mass of the patch (the mass transformed into an equivalent mass which is placed at the microcantilever's free end) is smaller when the patch gets closer to the fixed end, which produces higher resonant frequencies.

3.6   Resonant Microcantilever Arrays

Microcantilevers of identical cross sections and various lengths can be used in a resonant array formation as tools of detecting a range of bending resonant frequencies. In Chap. 4, similar arrays constructed of microbridges are presented, and more details are provided with respect to the precision of this spatial and frequency-domain discretization process. Figures 3.37 and 3.38 show two microcantilever architectures with members having rectangular and circular cross sections, respectively.

It is of interest here to determine the lengths of the component microcantilevers in such a way that each microcantilever matches a bending resonant frequency in a prescribed range. It can be shown that for a constant rectangular cross-section member, the length is related to a frequency f as

$$l = 0.405 \sqrt[4]{\frac{E t^2}{\rho f^2}} \qquad (3.179)$$

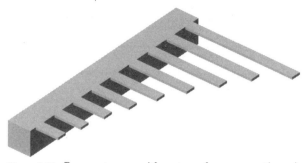

Figure 3.37 Resonant array with rectangular cross-section microcantilevers.

Figure 3.38 Resonant array with circular cross-section microcantilevers.

whereas for a circular cross-section microcantilever, the length-frequency relationship is

$$l = 0.377\sqrt[4]{\frac{Ed^2}{\rho f^2}} \qquad (3.180)$$

The plot of Fig. 3.39 shows the length as a function of frequency, according to Eq. (3.179), for polysilicon with E = 150 MPa, and ρ = 2300 kg/m^3 and when the thickness t = 0.5 μm.

As the figure indicates, the length decreases nonlinearly with the resonant frequency increasing. If a given distance needs to contain several microcantilevers, then the lengths of the microbridge network need to follow the profile of the plot shown in Fig. 3.39. A more detailed

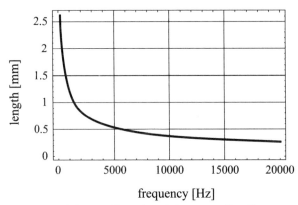

frequency [Hz]

Figure 3.39   Microcantilever length in terms of bending resonant frequency.

discussion of this topic is included in the presentation of resonant microbridge network, in Chap. 4.

References

1. J. M. Paros, and L. Weisbord, How to design flexure hinges, *Machine Design*, November 1965, pp. 151–156.

2. T. S. Smith, V. G. Badami, J. S. Dale, and Y. Xu, Elliptical flexure hinges, *Review of Scientific Instruments*, **68**(3), 1997, pp. 1474–1483.

3. N. Lobontiu, J. S. N. Paine, E. Garcia, and M. Goldfarb, Corner-filleted flexure hinges, *ASME Journal of Mechanical Design*, **123**, 2001, pp. 346–352.

4. N. Lobontiu, *Compliant Mechanisms: Design of Flexure Hinges*, CRC Press, Boca Raton, Fla., 2002.

5. N. Lobontiu, and E. Garcia, *Mechanics of Microelectromechanical Systems*, Kluwer Academic Press, New York, 2004.

6. R. Raiteri, M. Grattarola, H.-J. Butt, and P. Skladal, Micromechanical cantilever-based biosensors, *Sensors and Actuators B*, **79**, 2001, pp. 115–126.

7. B. Ilic, D. Czaplewski, M. Zalatudinov, and H. G. Craighead, Single cell detection with micromechanical oscillators, *Journal of Vacuum Science Technology*, **19**(6), 2001, pp. 2825–2828.

8. M. Sato, B. E. Hubbard, A. J. Sieves, B. Ilic, D. A. Czaplewski, and H. G. Craighead, Observation of locked intrinsic localized vibrational modes in a micromechanical oscillator array, *Physical Review Letters*, **90**(4), 2003, pp. 1–4.

9. B. Ilic, Y. Yang, and H. G. Craighead, Virus detection using nanoelectromechanical devices, *Applied Physics Letters*, **85**(13), 2004, pp. 2604–2606.

10. M. Zalatudinov, B. Ilic, D. Czaplewski, A. Zehnder, H. G. Craighead, and J. M. Parpia, Frequency-tunable micromechanical oscillator, *Applied Physics Letters*, **77**(20), 2000, pp. 2825–2828.

11. C. L. Britton, R. L. Jones, P. I. Oden, Z. Hu, R. J. Warmack, S. F. Smith, W. L. Bryan, and J. M. Rochelle, Multiple-input microcantilever sensors, *Ultramicroscopy*, **82**, 2000, pp. 17–21.

12. W. P. King, T. W. Kenny, K. E. Goodson, G. L. Cross, M. Despont, U. T. Durig, H. Rothuizen, G. Binning, and P. Vettiger, Design of atomic force microscope cantilevers for combined thermomechanical writing and thermal reading in array operation, *Journal of Microelectromechanical Systems,* **11**(6), 2002, pp. 765–774.

13. C. A. Peterson, R. K. Workman, X. Yao, J. P. Hunt, and D. Sarid, V-shaped metallic-wire cantilevers for combined atomic force microscopy and Fowler-Nordheim imaging, *Nanotechnology*, **9**, 1998, pp. 331–336.

14. S. Morita, R. Wiesendanger, and E. Meyer, *Noncontact Atomic Force Microscopy*, Imperial College Press, London, 1999.

15. W. van de Water, and J. Molenaar, Dynamics of vibrating atomic force microscopy, *Nanotechnology*, **11**, 2000, pp. 192–199.

16. R. P. Ried, J. Mamin, D. D. Terris, L.-S. Fan, and D. Rugar, 6-MHz 2-N/m piezoresistive atomic-force microscope cantilevers with INCISIVE tips, *Journal of Microelectromechanical Systems,* **6**(4), 1997, pp. 294–302.

17. N. Lobontiu and E. Garcia, Two microcantilever designs: modeling for static deflection and modal analysis, *Journal of Microelectromechanical Systems*, **13**(1), 2004, pp. 41–50.

18. E. Garcia, N. Lobontiu, and Y. Nam, Tuning the static and modal responses of microcantilevers through lumped-parameter model-based design, *2003 ASME IMECE Congress,* Washington, 2003.

19. E. Garcia, N. Lobontiu, Y. Nam, B. Ilic, and T. Reissman, Shape optimization of microcantilevers for mass variation detection and AFM applications, SPIE04 International Conference, San Diego, March 2004.

Microbridges: Lumped-Parameter Modeling and Design

4.1 Introduction

This chapter will analyze microbridges, which are fixed-fixed members, as shown in Fig. 4.1. Constructively, a microbridge might be identical to a microcantilever (or a microhinge), except for the end boundary conditions, although specific designs can be utilized for either category.

Bridges are mainly implemented in micro- and nanosensing and radio-frequency (RF) applications. Fabrication advances that permit size reduction of bridge resonators in the nanometer realm substantially improve the performance of devices that are designed to capture the effects of extraneous mass attachment. Attogram (10^{-15} g) quantities deposited on chemically prepared mechanical nanooscillators can be detected through shifts in the resonant frequencies. Simple doubly clamped beams and paddle bridges have been utilized to monitor various processes of mass addition by Ilic et al.[1] Sekaric et al.,[2,3] or Evoy et al.[4] among others. Such nanodevices perform with sensitivities in the 10^{-19} g/Hz domain and are capable of sensing deposition of substances at the cellular level. More details on mass addition detection by means of micro- and nanoresonators are given in Chap. 6.

One of the smallest NEMS oscillators reported, by Husain et al.,[5] is a nanowire only 1.3 μm long and 43 nm in diameter with a resonant frequency of more than 100 MHz. Microbridge resonators have also been researched from other angles to address topics such as electromechanical frequency tuning by Syms,[6] mechanical optimization for

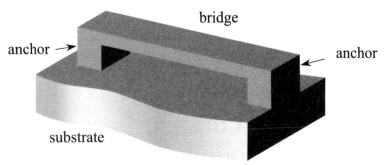

Figure 4.1   Microbridge of constant rectangular cross-section.

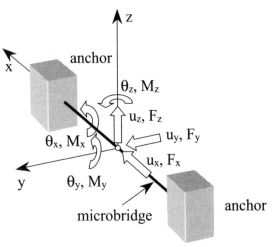

Figure 4.2   Generic microbridge with degrees of freedom at its centroid.

chemical sensing purposes by Dufour and Fadel,[7] nonlinear behavior above critical points (anharmonic motions) by Ayela and Fournier,[8] or torsional resonance by Plotz et al.[9]

Another class of microbridge applications targets the RF domain with MEMS devices such as capacitive switches, filters, or variable/tunable capacitors as reported by Yao et al.;[10] Peroulis, Pacheco, and Katehi;[11] or Park et al.;[12] to cite just a very small sample of the extended literature that has been dedicated to this topic. More details, based on a wide variety of studied examples of RF MEMS applications, are given in Rebeiz[13] whereas Madou[14] presents several micoresonator implementations, including the microbridge configuration that is the subject of this chapter.

The generic configuration sketched in Fig. 4.2 highlights the 6 degrees of freedom (three translations and three rotations) which are

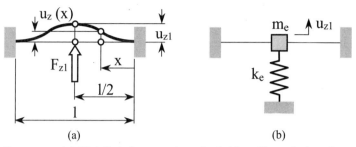

Figure 4.3 (*a*) Distributed-parameter microbridge; (*b*) equivalent lumped-parameter microbridge.

associated with the microbridge centroid. In general, bending and torsion are of interest in these micromembers, and therefore the axial resonant characteristics are not treated here. Moreover, for thin-film microcantilevers, such as those produced by surface micromachining, only bending about the sensitive axis is functionally relevant (the y axis in the sketch of Fig. 4.2). As a consequence, the focus in this chapter falls on deriving and characterizing the lumped-parameter resonant properties (through stiffness and inertia fractions) connected to bending about the sensitive y axis and torsion about the longitudinal x axis.

Microbridge configurations having constant cross section, of either homogeneous or multimorphic (sandwich) structure are studied first, followed by designs of variable cross sections, many of which are novel.

4.2   Microbridges of Constant Cross Section

Microbridges or fixed-fixed beams having constant cross sections are now analyzed by following a path similar to the one that has been detailed for microcantilevers. While the axial resonant characteristic might be important in characterizing the microcantilevers, as the resonant frequency can be on occasion close to the resonant frequencies in bending and torsion, for microbridges—especially for shorter ones — the axial resonant frequency is considerably larger than the other two relevant frequencies and therefore is not of paramount importance. As a consequence, the bending and torsional resonant frequencies are determined here for constant rectangular cross-section microbridges.

4.2.1   Bending resonant frequency

The lumping of stiffness and inertia is performed at the microbridge midspan so that the real, distributed-parameter system of Fig. 4.3*a* will be transformed to the equivalent lumped-parameter spring-mass system of Fig. 4.3*b*.

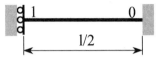

Figure 4.4   Half-model of a microbridge.

Similar to the modeling approach of microhinges and micro-cantilevers, both relatively long and short configurations are studied.

Long microbridges. Because of the geometric and load symmetry, the lumped-parameter stiffness and inertia can be determined by analyzing just half of the microbridge subject to the boundary conditions illustrated in Fig. 4.4. As known from the mechanics of materials, the stiffness at point 1 (and which is equal to one-half the stiffness of the full model) is

$$k_{b,e} = \frac{96EI_y}{l^3} \tag{4.1}$$

The effective mass, which needs to be placed at the guided end of the beam in Fig. 4.4, can be assessed by means of a distribution function that relates the deflection at a generic point (located at an abscissa x measured from the guided end, for instance) to the maximum deflection (at the guided end). It can be shown that this bending deflection distribution function is

$$f_b(x) = \frac{(l - 2x)^2(l + 4x)}{l^3} \tag{4.2}$$

When x is measured from the fixed end of the beam sketched in Fig. 4.4, the distribution function is expressed as:

$$f_b(x) = 4x^2(3l - 4x)/l^3 \tag{4.3}$$

According to Rayleigh's principle, as detailed in previous chapters and as shown by Timoshenko,[15] for instance, the effective mass that is dynamically equivalent to the distributed inertia of the half-length microbridge undergoing free bending vibrations is determined by equating the kinetic energy of the equivalent, lumped-parameter inertia to that of the distributed-parameter (real) system. In doing so, the effective mass is calculated by using either of the distribution functions given in Eqs. (4.2) and (4.3)

$$m_{b,e} = \rho A \int_0^{l/2} f_b^2(x)\, dx = \frac{13}{70m} \qquad (4.4)$$

with m being the total mass of the microbridge. The corresponding resonant frequency becomes

$$\omega_{b,e} = 22.736 \sqrt{\frac{EI_y}{ml^3}} \qquad (4.5)$$

Compared to the exact value of

$$\omega_b = 22.373 \sqrt{\frac{EI_y}{ml^3}} \qquad (4.6)$$

which is the solution to a partial differential equation, the approximate value of Eq. (4.5), which is higher, introduces a relative error of only 1.62 percent.

For a full-length microbridge model, the bending stiffness that is associated with the midspan is

$$k_{b,e} = \frac{192EI_y}{l^3} \qquad (4.7)$$

To find the effective mass, which needs to be placed at the midspan of the full-length microbridge, a method with initially unknown coefficients can be applied as shown in the following. The distribution function which relates the deflection of a generic point on the microbridge (at an abscissa x measured from one of the fixed ends) to the maximum deflection (at the midspan) can be found by assuming the following form of the deflection:

$$u_z(x) = a + bx + cx^2 + dx^3 + ex^4 \qquad (4.8)$$

The reason for choosing the particular polynomial of Eq. (4.8) with five unknown coefficients is that five separate boundary condition equations are available, namely, zero deflections at both ends and zero slopes at the ends and at the midpoint. The slope is the x-dependent derivative of the deflection, and therefore

$$\theta_y(x) = \frac{du_z(x)}{dx} = b + 2cx + 3dx^2 + 4ex^3 \qquad (4.9)$$

By using the five boundary condition equations

$$u_z(0) = u_z(l) = 0 \qquad \theta_y(0) = \theta_y\left(\frac{l}{2}\right) = \theta_y(l) = 0 \qquad u_z\left(\frac{l}{2}\right) = u_z \quad (4.10)$$

the five unknown coefficients can be found, and the following ratio, which is also the bending distribution function, can be formulated:

$$f_b(x) = \frac{u_z(x)}{u_z} = 16\frac{x^2}{l^2}\left(1 - \frac{x}{l}\right)^2 \tag{4.11}$$

The effective mass corresponding to the entire microbridge undergoing free bending vibrations is therefore

$$m_{b,e} = \rho A \int_0^l [f_b(x)]^2\, dx = \frac{128}{315m} \tag{4.12}$$

where, again, m is the total mass of the microbridge. By using Eqs. (4.7) and (4.12), the bending-related resonant frequency is

$$\omega_{b,e} = 21.737\sqrt{\frac{EI_y}{ml^3}} \tag{4.13}$$

It can be seen that, compared to the exact value of the resonant frequency in Eq. (4.6), the approximate value of Eq. (4.13) is lower, and the error between the two values is approximately 2.8 percent, which is still an acceptable figure.

Short microbridges.   In the case of relatively short microbridges where shearing effects are taken into account, the shear-dependent bending Eqs. (2.72) defining the half-length model are applied again. It can be shown that the angle θ_y is given by

$$\theta_y(x) = \frac{Fx(l/2 - x)}{2EI_y} \tag{4.14}$$

which resulted from the first of bending Eqs. (2.72). By substituting the expression of $\theta_y(x)$ of Eq. (4.14) into the second of bending Eqs. (2.72), the following distribution function is obtained:

$$f_b^{\text{sh}}(x) = (l - 2x)\frac{48\kappa EI_y + GA(l - 2x)(l + 4x)}{l(48\kappa EI_y + GAl^2)} \tag{4.15}$$

which connects the deflection $u_z(x)$ to the tip deflection u_z, according to Eq. (2.62). The tip deflection is

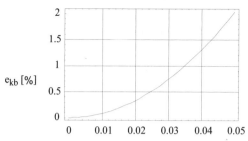

Figure 4.5   Relative errors between long and short microbridges in terms of stiffness.

$$u_z = u_z(0) = \frac{Fl[\kappa/(GA) + l^2/(48EI_y)]}{2} \qquad (4.16)$$

The stiffness of the half-model is

$$k_{b,e}^{\text{sh}} = \frac{F}{u_1} = \frac{96EGI_yA}{l(48\kappa EI_y + GAl^2)} \qquad (4.17)$$

Example: Compare the stiffness of a long configuration to that of a similar short configuration by analyzing one-half the length of a microbridge of constant cross section (defined by w and t) and length l.

If the error function is formulated as

$$e_{kb} = \frac{k_{b,e} - k_{b,e}^{\text{sh}}}{k_{b,e}} 100 \qquad (4.18)$$

the following equation is obtained by using Eqs. (4.13) and (4.17):

$$e_{kb} = \frac{1}{1 + Gl^2/(4\kappa Et^2)} 100 = \frac{100\alpha}{0.12 + \alpha^2} \qquad (4.19)$$

where it has been considered that

$$\alpha = \frac{t}{l} \qquad (4.20)$$

and that $\kappa = \frac{5}{6}$ (for a rectangular cross section) and $\mu = 0.25$ (Poisson's ratio for polysilicon) and that E and G are related according to Eq. (2.70). The relative error function of Eq. (4.19) is plotted in Fig. 4.5, which shows that the maximum deviation between the long- and short-configuration model predictions is approximately 2 percent.

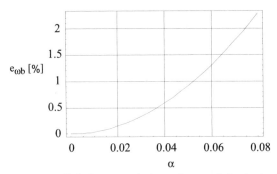

Figure 4.6 Relative errors between long and short microbridges in terms of the bending resonant frequency.

The effective mass of the half-model is determined by using the distribution function of Eq. (4.15), and its equation is

$$m_{b,e}^{sh} = \frac{26{,}880\kappa^2 E^2 I_y^2 + 1176\kappa EGI_y Al^2 + 13G^2 A^2 l^4}{\left[70(GAl^2 + 48\kappa EI_y)^2\right]} \; m \qquad (4.21)$$

and therefore the lumped-parameter resonant frequency becomes

$$\omega_b^{sh} = 8\sqrt{105}\sqrt{\frac{EGI_y A(48\kappa EI_y + GAl^2)}{l(26{,}880\kappa^2 E^2 I_y^2 + 1176\kappa EGI_y Al^2 + 13G^2 A^2 l^4)m}} \qquad (4.22)$$

It can again be shown that in the case where the shearing effects are neglected (zero shearing force in Timoshenko's model), Eqs. (4.17), (4.21), and (4.22) reduce to Eqs. (4.7), (4.12), and (4.13), respectively, corresponding to the Euler-Bernoulli model.

Example: Study the influence of the shearing effects on the bending resonant frequency of a constant rectangular cross-section microbridge that is constructed of a material with a density of $\rho = 3000$ kg/m³.

By using the data of the previous example, a similar comparison is performed here between the resonant bending frequency of a microbridge with and without considering the shearing effects, by means of the frequency ratio

$$e_{\omega b} = \frac{\omega_{b,e} - \omega_{b,e}^{sh}}{\omega_{b,e}} 100 \qquad (4.23)$$

which can be again expressed in terms of just the nondimensional parameter α of Eq. (4.20).

As illustrated in Fig. 4.6, the errors in calculating the bending resonant frequency according to the long-beam model as opposed to the short-beam model are less than 2 percent; moreover, for bridges where the thickness-to-length ratio is small, these errors are smaller and therefore almost negligible.

4.2.2 Torsion resonant frequency

The torsional resonant frequency can be determined by finding the lumped-parameter stiffness and inertia for a half-length microbridge and for the full-length microbridge, respectively.

For the half-length microbridge, according to the model sketched in Fig. 4.4, a moment applied about the longitudinal (x) axis at the guided end (which, as far as torsion is concerned, is considered free) produces torsion of the bar, and it can simply be shown that the torsional stiffness of that segment (the ratio of the applied moment to the resulting rotation angle) is

$$k_{t,e} = \frac{2GI_t}{l} \qquad (4.24)$$

The lumped-parameter mechanical moment of inertia of an equivalent rigid body which is placed at the guided end is determined by means of Rayleigh's principle again. The torsion-related distribution function is the ratio of the rotation angle at a generic point of abscissa x (measured from point 1 toward the right in Fig. 4.4) to the maximum rotation angle (at point 1 in the same figure) and is found to be

$$f_t(x) = 1 - \frac{2x}{l} \qquad (4.25)$$

As a consequence, the lumped-parameter mechanical moment of inertia becomes

$$J_{t,e} = \frac{J_t}{6} \qquad (4.26)$$

where J_t is the torsional mechanical moment of inertia of the full-length microbridge. By combining Eqs. (4.24) and (4.26), the torsional resonant frequency for a half-microbridge becomes

$$\omega_{t,e} = 3.46 \sqrt{\frac{GI_t}{lJ_t}} \qquad (4.27)$$

The same result should be found when the calculus is performed for the full-length microbridge, such as the one shown in Fig. 4.7.

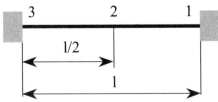

Figure 4.7   Full-length microbridge for torsional resonant frequency calculation.

The lumped-parameter stiffness (the ratio of the torsion moment applied at the midspan — point 2 — to the resulting angular deformation) is

$$k_{t,e} = \frac{4GI_t}{l} \qquad (4.28)$$

and this is twice the stiffness of the half-length microbridge [Eq. (4.24)], as expected.

The distribution function, which is needed to determine the lumped-parameter mechanical moment of inertia corresponding to point 2 in Fig. 4.7, and which is dynamically equivalent to the distributed mass of the full-length microbridge undergoing free torsional vibrations, is found by applying the method of unknown coefficients which has been presented for bending vibrations. The torsional angle at a generic point of abscissa x is sought of the following polynomial form:

$$\theta_x(x) = a + bx + cx^2 \qquad (4.29)$$

The three unknown coefficients in Eq. (4.29) are determined by using the following boundary conditions:

$$\theta_x(0) = \theta_x(l) = 0 \qquad \theta_x\!\left(\frac{l}{2}\right) = \theta_x \qquad (4.30)$$

As a result, the distribution function can be expressed as

$$f_t(x) = \frac{\theta_x(x)}{\theta_x} = \frac{4x(l-x)}{l^2} \qquad (4.31)$$

It can be seen that this function satisfies the expected conditions:

$$f_t(x) = \begin{cases} 0 & x = 0 \text{ and } x = l \\ 1 & x = \dfrac{l}{2} \end{cases} \qquad (4.32)$$

With the distribution function of Eq. (4.31), the lumped-parameter mechanical moment of inertia at the midspan is

$$J_{t,e} = \frac{8J_t}{15} \tag{4.33}$$

The torsional resonant frequency results from Eqs. (4.28) and (4.33), and its equation is

$$\omega_{t,e} = 2.74 \sqrt{\frac{GI_t}{lJ_t}} \tag{4.34}$$

It can be seen that the resonant frequency corresponding to the full-length microbridge is not identical to that yielded by the half-length model. It can be shown by using distributed-parameter modeling, and also as indicated by Rao,[16] that the exact torsional frequency is

$$\omega_t = \pi \sqrt{\frac{GI_t}{lJ_t}} \tag{4.35}$$

Comparison of the exact torsional frequency to those produced by the half- and full-length lumped-parameter models indicates that the lumped-parameter half-length model prediction overevaluates the exact value of the resonant frequency, whereas the full-length model result underevaluates the exact value, and this is similar to the situation corresponding to bending.

Example: Compare the bending resonant frequency to the torsional one by analyzing a rectangular constant-cross-section microbridge in terms of the defining geometry parameters.

The following resonant frequency ratio can be formed:

$$r\omega = \frac{\omega_{t,e}}{\omega_{b,e}} \tag{4.36}$$

By using the nondimensional parameters

$$c_w = \frac{w}{t} \qquad c_l = \frac{l}{t} \tag{4.37}$$

the frequency ratio of Eq. (4.36) is plotted in Fig. 4.8. The torsional resonant frequency is higher than the bending one, and the frequency ratio increases with increasing length-to-thickness parameter c_l and decreasing width-to-thickness ratio c_t.

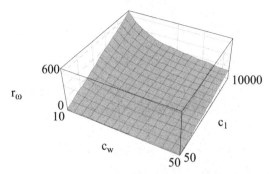

Figure 4.8   Torsion-to-bending resonant frequency ratio.

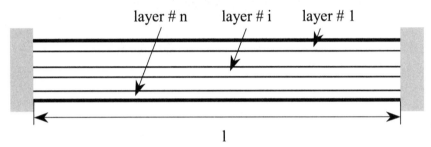

layer # n   layer # i   layer # 1

1

Figure 4.9   Side view of a multimorph microbridge.

4.3   Sandwiched Microbridges (Multimorphs)

The problems discussed for multimorph microcantilevers are also valid for multimorph microbridges (which are cantilevers fixed at both ends), and the relevant resonant frequencies are derived in the following for equal-length and dissimilar-length multimorph configurations. Because axial vibration is seldom used in microresonators of this kind, only torsional and bending resonant frequencies are calculated next, as was done previously.

4.3.1   Multimorph microbridges of equal-length layers

The side view of an equal-length multimorph microbridge is sketched in Fig. 4.9. The assumption is also made here that all layers have the same width w.

The bending resonant frequency is again determined by first finding the lumped-parameter stiffness and mass corresponding to the microbridge midpoint, in a manner similar to that used in finding the lumped-parameter properties of homogeneous (one-component) microbridges. It has been shown that the stiffness and mass of the half-microbridge are one-half the corresponding properties of the full microbridge (exactly for stiffnesses and approximately for mass fractions), which results in the natural frequencies predicted by the two models being approximately equal. It can simply be shown that the lumped-parameter stiffness for the half-microbridge (the midpoint is guided, as indicated in Fig. 4.4) is

$$k_{b,e} = \frac{96(EI_y)_e}{l^3} \tag{4.38}$$

with the equivalent bending rigidity $(EI_y)_e$ being calculated according to Eq. (3.148). The lumped mass for the half-microbridge is

$$m_{b,e} = \frac{13}{70} l w \sum_{i=1}^{n} (\rho_i A_i) \tag{4.39}$$

such that the bending natural frequency is

$$\omega_{b,e} = \sqrt{\frac{k_{b,e}}{m_{b,e}}} = 22.736 \frac{\sqrt{(EI_y)_e / (w \sum_{i=1}^{n} \rho_i t_i)}}{l^2} \tag{4.40}$$

Equation (4.40) reduces to the following for a bimorph (two-component) microbridge:

$$\omega_{b,e} = \frac{6.563}{l^2} \sqrt{\frac{E_1^2 t_1^4 + E_2^2 t_2^4 + 2E_1 E_2 t_1 t_2 (2t_1^2 + 3t_1 t_2 + 2t_2^2)}{(E_1 t_1 + E_2 t_2)(\rho_1 t_1 + \rho_2 t_2)}} \tag{4.41}$$

When $t_2 \to 0$, Eq. (4.41) simplifies to

$$\omega_b = 22.736 \sqrt{\frac{E_1 I_{y1}}{m_1 l^3}} \tag{4.42}$$

which is the equation providing the bending resonant frequency of a homogeneous (one-component) microbridge.

In torsion, the stiffness pertaining to the half-microbridge is expressed as

$$k_{t,e} = \frac{2(GI_t)_e}{l} \tag{4.43}$$

where the equivalent torsional rigidity is

$$(GI_t)_e = \sum_{i=1}^{n} G_i I_{ti} \tag{4.44}$$

The mechanical moment of inertia which is equivalent to the distributed inertia of the half multimorph microbridge can be calculated as

$$J_{t,e} = \frac{1}{6} wl \sum_{i=1}^{n} \rho_i t_i \frac{w^2 + t_i^2}{12} \tag{4.45}$$

which, according to the example analyzed in Chap. 3, ignored the terms in the individual inertias that were calculated in terms of an axis passing through the symmetry center of the compound cross section, and which were shown to be negligibly small. For a bimorph, Eq. (4.45) reduces to

$$J_{t,e} = \frac{1}{72} wl[\rho_1 t_1(t_1^2 + w^2) + \rho_2 t_2(t_2^2 + w^2)] \tag{4.46}$$

When $t_2 \to 0$, Eq. (4.46) further simplifies to

$$J_{t,e} = \frac{1}{72} m_1(w^2 + t_1^2) \tag{4.47}$$

which is the known relationship for a single-component, homogeneous bar.

The resonant frequency of a multimorph microbridge is found by combining Eqs. (4.43) and (4.45) in the form:

$$\omega_{t,e} = \sqrt{\frac{k_{t,e}}{J_{t,e}}} = \frac{(3.46/l)\sqrt{\sum_{i=1}^{n} G_i I_{ti}}}{\sqrt{w \sum_{i=1}^{n} \rho_i t_i(w^2 + t_i^2)}} \tag{4.48}$$

For a bimorph, Eq. (4.48) changes to

$$\omega_t = \frac{6.93}{l} \sqrt{\frac{G_1 t_1^3 + G_2 t_2^3}{\rho_1 t_1(t_1^2 + w^2) + \rho_2 t_2(t_2^2 + w^2)}} \tag{4.49}$$

Equation (4.49) simplifies to Eq. (4.27) when $t_2 \to 0$ (one-component microbridge).

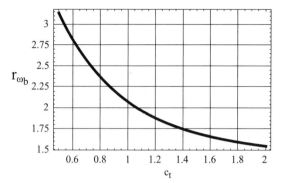

Figure 4.10   Bimorph-to-unimorph bending resonant frequency ratio.

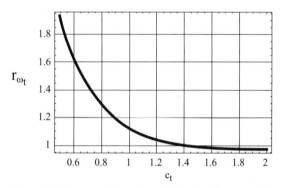

Figure 4.11   Bimorph-to-unimorph torsional resonant frequency ratio.

Example: Compare the resonant behavior in bending and torsional free vibrations of a unimorph microbridge to that of a bimorph microbridge. The unimorph and bimorph have identical lengths and widths. Known also: $E_1 = 160$ GPa, $G_1 = 65$ GPa, $E_2 = 170$ GPa, $G_2 = 70$ GPa, $\rho_1 = 2000$ kg/m^3, and $\rho_2 = 2300$ kg/m^3. Also use the baseline parameter values of $w = 50$ µm and $t_1 = 1$ µm.

The bimorph-to-unimorph bending resonant frequency ratio is formed by using Eqs. (4.41) and (4.6), and Fig. 4.10 displays the two-dimensional plot of this ratio as a function of the thickness parameter $c_t = t_2 / t_1$.

As Fig. 4.10 shows, the bending resonant frequency of the bimorph is always larger than that of the unimorph for the thickness parameter range selected in this example. Similarly, the bimorph-to-unimorph torsional resonant frequency ratio is formed by using Eqs. (4.49) and (4.35), and Fig. 4.11 is the two-dimensional plot of this ratio as a function of the same thickness parameter.

The differences between the torsional resonant frequency of the bimorph and that of the corresponding unimorph are less pronounced, and their values tend to be equal for thicker substrates (large c_t values).

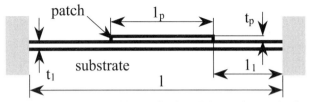

Figure 4.12   Side view of dissimilar-length bimorph microbridge.

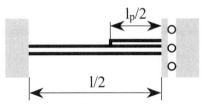

Figure 4.13   Half-model of dissimilar-length bimorph microbridge.

4.3.2 Multimorph microbridges of dissimilar-length layers

A dissimilar-length bimorph microbridge is shown in side view in Fig. 4.12. Constructively, the microbridge is similar to the bimorph microcantilever of Fig. 3.35. The particular case is only studied here where the patch layer is positioned symmetrically with respect to the midspan, and therefore $l_1 = (l - l_p)/2$. The bending and torsional resonant frequencies are determined for this configuration by again considering a half-length microbridge which is formed of two portions: one of length l_1 at the root and the other of length $l_p/2$ extending from the root region (see Fig. 4.13).

In bending, the half-length model is fixed at one end and guided at the opposite one, as shown in Fig. 4.13. The stiffness with respect to the midpoint of Fig. 4.12 — the guided end in Fig. 4.13 — is calculated by applying a force at that point on the half-model and by calculating the corresponding deflection by means of Castigliano's displacement theorem, for instance. The resulting stiffness is

$$k_{b,e} = \frac{96E_1 I_{y1}(EI_y)_e[l_p E_1 I_{y1} + (l - l_p)(EI_y)_e]}{[E_1^2 I_{y1}^2 l_p^4 + 2l_p(l - l_p)(2l^2 - l l_p + l_p^2)E_1 I_{y1}(EI_y)_e}$$
$$+(l - l_p)^4 (EI_y)_e^2]} \tag{4.50}$$

In Eq. (4.50) the rigidity $(EI_y)_e$ is calculated according to Eq. (3.148) over the length $l_p/2$ and by incorporating the substrate and patch elastic and geometric properties. Equation (4.50) reduces to

$$k_b = \frac{96 E_1 I_{y1}}{l^3} \tag{4.51}$$

when $l_p \to 0$ (no patch on the substrate).

The equivalent mass, which is placed at the guided end of the half-model patched bimorph of Fig. 4.13, is calculated as

$$m_{b,e} = \frac{13}{70} (m_1 + f_p m_p) \tag{4.52}$$

where m_1 and m_p are the masses of the substrate and patch, respectively, and

$$f_p = \frac{2.69 l^6 - 5.38 l^4 l_p^2 + 2.69 l^3 l_p^3 + 4.85 l^2 l_p^4 - 5.38 l l_p^5 + 1.54 l_p^6}{l^6} \tag{4.53}$$

When $m_p \to 0$, Eqs. (4.52) and (4.53) simplify to

$$m_b = \frac{13}{70} m_1 \tag{4.54}$$

which is the known relationship for a one-component, half-model microbridge. The bending resonant frequency as provided by the half-microbridge model is

$$\omega_{b,e} = \frac{22.736 \sqrt{E_1 I_{y1} (EI_y)_e [l_p E_1 I_{y1} + (l - l_p)(EI_y)_e]}}{\sqrt{\begin{array}{l} E_1^2 I_{y1}^2 l_p^4 + 2 l_p (l - l_p)(2l^2 - l l_p) \\ + l_p^2) E_1 I_{y1} (EI_y)_e + (l - l_p)^4 (EI_y)_e^2 \end{array}} \bigg| \sqrt{m_1 + f_p m_p}} \tag{4.55}$$

Obviously, Eq. (4.55) simplifies to Eq. (4.5) when $l_p \to 0$ and $m_p \to 0$, which is the known resonant frequency equation for a one-component (homogeneous) microbridge. It can also be checked that when $l_p = l$, Eq. (4.55) changes to Eq. (4.42), which expresses the bending resonant frequency of an equal-length layer bimorph.

In torsion, the stiffness of the half-model patched microbridge of Fig. 4.13 is calculated by applying a torsional moment at the guided end and by calculating the corresponding rotation angle at the same point about the longitudinal direction. Under the assumption of very thin layers for both the substrate and the patch, this stiffness is

$$k_{t,e} = \frac{2}{l_p / (GI_t)_e + (l - l_p) / (G_1 I_{t1})} \tag{4.56}$$

where the equivalent torsional rigidity is expressed as

$$(GI_t)_e = \frac{w(t_1^3 + t_p^3)}{3} \tag{4.57}$$

In the end and by way of Eqs. (4.56) and (4.57), the torsional stiffness becomes

$$k_{t,e} = \frac{2G_1 t_1^3 w(G_1 t_1^3 + G_p t_p^3)}{3[G_1 l t_1^3 + G_p (l - l_p) t_p^3]} \tag{4.58}$$

When $l_p \to 0$, the stiffness of Eq. (4.58) reduces to

$$k_t = \frac{2Gwt_1^3}{3l} \tag{4.59}$$

which is indeed the torsional stiffness of the half-length patched microbridge.

The equivalent inertia fraction (mechanical moment of inertia) which needs to be placed at the guided end of the half-model microbridge of Fig. 4.13 is again calculated by means of Rayleigh's principle and by assuming the following velocity distribution:

$$\theta_x(x) = \left(1 - \frac{2x}{l}\right)\theta_x \tag{4.60}$$

By using this procedure, the equivalent mechanical moment of inertia becomes

$$J_{t,e} = \frac{w}{72} \left[\frac{\rho_p l_p t_p (w^2 + t_p^2)(3l^2 - 3l l_p + l_p^2)}{l^2} \right] + \rho_1 l t_1 (w^2 + t_1^2) \tag{4.61}$$

When $l_p \to 0$ and $t_p \to 0$ (no patch, just the substrate), Eq. (4.61) reduces to

$$J_{t,e} = \frac{J_t}{6} = \frac{m_1(w^2 + t_1^2)}{72} \tag{4.62}$$

which is the equation giving the equivalent inertia for the half-length model of a single-component microbridge. The torsion-related resonant frequency is obtained by combining Eqs. (4.58) and (4.61), namely,

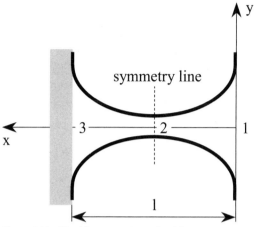

Figure 4.14 Double-symmetry microhinge.

$$\omega_{t,e} = \frac{6.93t_1\sqrt{G_1t_1(G_1t_1^3 + G_pt_p^3)/[G_1lt_1^3 + G_p(l - l_p)t_p^3]}}{\sqrt{\rho_pl_pt_p(w^2 + t_p^2)(3l^2 - 3ll_p + l_p^2)/l^2 + \rho_1lt_1(w^2 + t_1^2)}} \qquad (4.63)$$

4.4 Microbridges of Variable Cross Section

This is the largest section of this chapter, and we study the lumped-parameter resonant properties of variable-cross-section microbridges. Specific designs are presented that belong to one of the following categories: single-profile, two-segment, and three-segment microbridge configurations. The relevant stiffness characteristics of variable-cross-section microbridges are defined in terms of compliances of simpler (basic) segments of various shapes, whose properties have been derived in Chap. 2 and have also been studied by Lobontiu[17] and more recently by Lobontiu and Garcia.[18]

4.4.1 Compliance transform

Two compliance transforms were defined in Chap. 3 which allowed us to express the three bending compliances with respect to arbitrarily translated reference frames and reversed reference frames. Another compliance transform is presented here that enables us to calculate the bending compliances of a half-microhinge in terms of the compliances defining a full-length hinge, which possesses a transverse symmetry axis, as sketched in Fig. 4.14. The requirement is also imposed that the profile of the microhinge must be defined by a single curve.

This transform is mainly useful when we are carrying out stiffness calculations for microbridges which are formed of a single geometric curve (profile) and which possess a transverse symmetry axis. The subscript f is used here to denote the full-length microhinge whereas for the half portion between points 2 and 3 in Fig. 4.14, no additional subscript notation is utilized. The following equations can be written in terms of bending compliances:

$$C_{l,f} = \int_0^l \frac{x^2\,dx}{EI_y} = \int_0^{l/2} \frac{x^2\,dx}{EI_y} + \int_{l/2}^l \frac{x^2\,dx}{EI_y} = \frac{l^2 C_r}{2} + 2C_l$$

$$C_{c,f} = \int_0^l \frac{x\,dx}{EI_y} = \int_0^{l/2} \frac{x\,dx}{EI_y} + \int_{l/2}^l \frac{x\,dx}{EI_y} = l C_r \tag{4.64}$$

$$C_{r,f} = \frac{2C_c}{l}$$

Equations (4.64) have been derived by applying the two compliance transforms of Chap. 3. By solving the first two of Eqs. (4.64), the two compliances that correspond to one-half the microhinge of Fig. 4.14 are determined to be

$$C_l = \frac{2C_{l,f} - l C_{c,f}}{4} \qquad C_r = \frac{C_{c,f}}{l} \tag{4.65}$$

It can also be shown that the torsional compliance of the segment 2-3 can be expressed in terms of the compliance of the full-length micro-hinge as

$$C_t = \frac{C_{t,f}}{2} \tag{4.66}$$

4.4.2 Generic formulation for single-profile (basic shape) microbridges

As was the case with constant-cross-section microbridges, only a half-model of a variable-cross-section microbridge is analyzed, provided there is transverse symmetry of the design, which means that one-half of the microbridge is mirrored-identical to the other half. Figure 3.7, which was used for symmetric microcantilevers, is also valid for this derivation, and the generic half-bridge model of Fig. 4.4 is as well. The assumption is made with this model that the microbridge width is variable, whereas the thickness is constant. By applying a force F_z at the guided end in Fig. 4.4, it can be shown that the bending stiffness related to that point is

$$k_{b,e} = \frac{F_z}{u_z} = \frac{1}{C_l - C_c^2/C_r} \qquad (4.67)$$

where the compliances defining these stiffness are calculated as

$$C_l = \frac{12}{E} \int_0^{l/2} \left\{ x^2 / [t^3 w(x)] \right\} dx$$

$$C_c = \frac{12}{E} \int_0^{l/2} \left\{ x / [t^3 w(x)] \right\} dx \qquad (4.68)$$

$$C_r = \frac{12}{E} \int_0^{l/2} dx / [t^3 w(x)]$$

A check has been performed on Eq. (4.67) to verify whether Eq. (4.1) is obtained when the rectangular cross-section is constant, and this was indeed the case.

For short microcantilevers, where shearing effects are taken into consideration and the Timoshenko model is utilized, the bending stiffness is expressed as

$$k_{b,e}^{sh} = \frac{1}{C_l^{sh} - C_c^2/C_r} \qquad (4.69)$$

where the direct-bending linear shear–dependent compliance is expressed as in Eq. (2.28) with the aid of the axial compliance equation:

$$C_a = \frac{1}{E} \int_0^{l/2} dx / [t w(x)] \qquad (4.70)$$

It can be checked again that for a constant rectangular cross-section microbridge, Eq. (4.69) reduces to Eq. (4.17), and this proves the validity of the generic formulation.

The lumped mass which needs to be placed at the guided end of the half-microbridge in Fig. 4.4 and which is dynamically equivalent to the distributed inertia of the half-length microbidge undergoing free bending vibrations can be calculated by means of the corresponding distribution function. This bending-related distribution function is determined in the usual manner as the ratio of the deflection produced at a generic point on the interval [0, $l/2$] to the maximum deflection (at the guided end) under the action of a force F_z applied at the guided end. It can be shown that for a variable cross section, the bending-related distribution function is

$$f_b(x) = \frac{C_l(x) - (x + C_c/C_r)C_c(x) + xC_c/C_rC_r(x)}{C_l - C_c^2/C_r} \tag{4.71}$$

where the abscissa-dependent compliances in the numerator are defined as

$$C_l(x) = \frac{12}{E} \int_x^{l/2} \left\{ x^2 / [t^3 w(x)] \right\} dx$$

$$C_c(x) = \frac{12}{E} \int_x^{l/2} \left\{ x / [t^3 w(x)] \right\} dx \tag{4.72}$$

$$C_r(x) = \frac{12}{E} \int_x^{l/2} dx / [t^3 w(x)]$$

The distribution function of Eq. (4.71) with the associated Eqs. (4.72) results in

$$f_b(x) = \frac{(l - 2x)^2(l + 4x)}{l^3} \tag{4.73}$$

when the rectangular cross section is constant. Equation (4.73) is actually identical to Eq. (4.2) which directly expresses the bending-related distribution function for a half-length microbridge.

For short microbridge configurations, Eq. (4.71) becomes

$$f_b^{\text{sh}}(x) = \frac{C_l^{\text{sh}}(x) - (x + C_c/C_r)C_c(x) + xC_c/C_rC_r(x)}{C_l^{\text{sh}} - C_c^2/C_r} \tag{4.74}$$

For a constant-cross-section microbridge, the generic Eq. (4.74) simplifies to

$$f_b^{\text{sh}}(x) = \frac{[48\kappa EI_y + GA(l - 2x)(l + 4x)](l - 2x)}{l(48\kappa EI_y + GAl^2)} \tag{4.75}$$

where $\qquad I_y = \dfrac{w(x)t^3}{12} \qquad A = w(x)t$ $\qquad\qquad$ (4.76)

The lumped-parameter mass, which is dynamically equivalent to the distributed inertia of the half-microbridge undergoing free bending vibrations, is calculated as

$$m_{b,e} = \rho t \int_0^{l/2} f_b^2(x)w(x)\,dx \tag{4.77}$$

When the rectangular cross section is constant, Eq. (4.77) reduces to Eq. (4.12), which qualifies the bending inertia fraction of a constant rectangular cross-section half-bridge. For short microbridges, $f_b(x)$ is substituted by $f_b^{\text{sh}}(x)$ in Eq. (4.77).

The lumped-parameter resonant frequency which is associated with free bending vibrations combines the stiffness of Eq. (4.67) and mass of Eq. (4.77) for a relatively long microbridge in the form:

$$\omega_{b,e} = \frac{1}{\sqrt{\rho t(C_l - C_c^2 \big/ C_r)\displaystyle\int_0^{l/2} f_b^2(x)w(x)\,dx}} \tag{4.78}$$

For short microbridge configurations, C_l^{sh} needs to be used instead of C_l and $f_b^{\text{sh}}(x)$ instead of $f_b(x)$ in Eq. (4.78).

In torsion, the stiffness of the half-microbridge is simply

$$k_{t,e} = \frac{1}{C_t} \tag{4.79}$$

where $\qquad C_t = \dfrac{1}{G}\displaystyle\int_0^{l/2} \dfrac{dx}{I_t}$ $\qquad\qquad$ (4.80)

The lumped mechanical moment of inertia which is equivalent to the distributed inertia of the torsionally vibrating half-microbridge is

$$J_{t,e} = \frac{\rho t}{12} \times \int_0^{l/2} f_t^2(x)w(x)[w(x)^2 + t^2]\,dx \tag{4.81}$$

where the distribution function is

$$f_t(x) = \frac{C_t(x)}{C_t} \qquad (4.82)$$

with
$$C_t(x) = \frac{1}{G} \times \int_x^{l/2} (dx / I_t) \qquad (4.83)$$

where the torsional moment of inertia is

$$I_t = \frac{w(x)t^3}{3} \qquad (4.84)$$

The lumped-parameter torsional resonant frequency is

$$\omega_{t,e} = \frac{3.46}{\sqrt{\rho\, tC_t \int_0^{l/2} f_t^2(x)w(x)[w(x)^2 + t^2]\, dx}} \qquad (4.85)$$

A right elliptic microbridge is used as an example to illustrate the generic formulation that has been developed here. Structurally, the right elliptic microbridge is identical to the right elliptic microhinge studied in Chap. 3 and sketched in Fig. 3.22.

Example: Find the lumped-parameter bending-related resonant frequency of an elliptic microbridge by using the single-profile microbridge model.

The bending stiffness which is associated with the midpoint of the microbridge is

$$k_{b,e} = \frac{4Eb^3t^3}{3a^3\big\{2(4+\pi)b^2 + 4(1+\pi)bw_1 + \pi w_1^2}$$
$$+4(2b+w_1)\sqrt{w_1(4b+w_1)}\arctan\sqrt{1+4b/w_1} \qquad (4.86)$$
$$-2[(2b+w_1)\ln(1+2b/w_1)-2b]^2$$
$$/[2(2b+w_1)\arctan\sqrt{1+4b/w_1}/\sqrt{w_1(4b+w_1)}-\pi/2]\big\}$$

The effective mass is

$$m_{b,e} = \rho t a(14.28b + 9.18w_1) \qquad (4.87)$$

and the bending-related resonant frequency is simply determined by combining Eqs. (4.86) and (4.87).

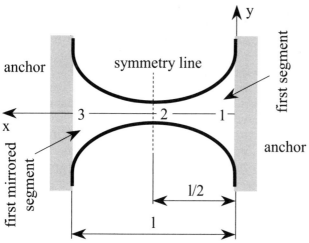

Figure 4.15   Top view of microbridge formed of identical and mirrored compliant segments.

4.4.3 Serially compounded microbridges

Microbridges of more complex geometric configurations can be formed by serially combining basic shapes, similar to the procedure that detailed in Chap. 3 where several compound microcantilever and microhinge designs have been studied. Two generic cases of serially compound microbridges with associated applications that are implemented in MEMS design are studied in this section, namely, two-segment and three-segment configurations.

Two-segment microbridges.   One class of serially compounded microbridges is formed of configurations containing two identical basic units that are placed in mirror with respect to the midpoint of the resulting microbridge, as shown in Fig. 4.15. This category resembles the single-profile microbridge class which has just been studied. While the single-profile microbridges are defined by one curve only over the whole length, the two-segment designs, although preserving the transverse symmetry property, can be defined by several curves over one half. As a consequence, the two-segment fixed-fixed flexible component has transverse symmetry, and the structurally identical case of a generic fixed-free member was analyzed in Chap. 3 based on the sketch of Fig. 3.7, which is reproduced here in Fig. 4.15 to highlight the changes in boundary conditions.

As shown in the following, the bending and torsional resonant frequencies of the entire structure can be expressed in terms of compliances and inertia-related properties that define only one-half the component, namely, the second (left) segment, if the reference frame origin is chosen at the right fixed end, as shown in Fig. 4.15. A similar approach was taken, as already mentioned, in analyzing two-segment microhinges in Chap. 3, where only the properties of one basic segment of the two were utilized. The subscripts (2) and (1) have been employed to denote the left and the right segments, respectively. To simplify notation, no subscript is used here, but all compliances, for instance, are those of the second mirrored segment of Fig. 4.15, taken with respect to point 2 (which is assumed free) when point 3 is fixed. In the end, the generic formulation produces the bending and torsional resonant frequencies that are associated with the midpoint (point 2) of the two-segment microbridge of Fig. 4.15.

Bending resonant frequency. The lumped-parameter bending-related stiffness and effective mass are determined here for the entire flexible structure of Fig. 4.15. To express the lumped-parameter stiffness at point 2, which is the ratio of a force applied at that point about the z direction to the resulting out-of-the-plane deflection, the following two steps have to be undertaken:

• Calculation of the force and moment reactions at fixed point 1 when a force is applied about the z axis at midpoint 2 in terms of that force by using compliances

• Calculation of the deflection produced at midpoint 2 under the action of the force applied at that point

Both steps need to utilize the compliance transforms presented in Chap. 3 which enable us to express compliances in terms of switched and arbitrarily translated reference frames.

The lumped-parameter bending stiffness that is associated with midpoint 2 can be expressed as

$$k_{b,e} = \frac{1}{[1 + 2(c_1 - 1)c_1]C_l + (c_2 - c_1 l / 2)[2C_c + (2c_2 - c_1 l)C_r]} \quad (4.88)$$

The constants c_1 and c_2 which enter Eq. (4.88) are

$$c_1 = \frac{a_{12}b_1 + a_{11}b_2}{a_{11}a_{22} - a_{12}^2} \qquad c_2 = \frac{a_{22}b_1 + a_{12}b_2}{a_{11}a_{22} - a_{12}^2} \quad (4.89)$$

where $\qquad a_{11} = C_r' + C_r'' \quad a_{12} = C_c' + C_c'' \quad a_{22} = C_l' + C_l'' \quad (4.90)$

and $\qquad b_1 = \dfrac{l}{2C_r''} - C_c'' \quad b_2 = \dfrac{-l}{2C_c''} + C_l'' \quad (4.91)$

As a reminder, the prime in Eqs. (4.90) and (4.91) indicate compliances that are calculated for the second segment in Fig. 4.15 with respect to a switched reference frame (located at point 3), whereas the double prime shows compliances calculated for the same second segment with respect to a reference frame placed at point 1. Both compliance sets can be expressed in terms of the bending-related compliances of the second segment (regularly calculated with respect to the reference frame placed at point 2) according to the compliance transformations of Eqs. (3.2) and (3.10).

It should be emphasized that the generic Eq. (4.88) gives the lumped-parameter stiffness of the entire two-segment microbridge by using compliances that define one-half the microbridge, specifically one of the two identical segments, and which can be denoted by C_l, C_c, and C_r. A check has been performed of this generic calculation algorithm, by considering the two identical segments of constant rectangular section. By using the equations which define C_l, C_c, and C_r of a constant rectangular cross-section fixed-free segment of length $l/2$, Eq. (4.7) is obtained, which defines the midpoint stiffness of a constant rectangular cross-section microbridge of length l.

The lumped-parameter effective mass which needs to be placed at midpoint 2, and is dynamically equivalent to the distributed mass of the two-segment microbridge undergoing free out-of-the-plane bending vibrations, is calculated by

$$m_{b,e} = \rho t \int_0^l w(x) f_b^2(x)\, dx \tag{4.92}$$

where the bending distribution function $f_b(x)$ is expressed in Eq. (4.11) and the variable width $w(x)$ can be expressed as

$$w(x) = \begin{cases} w(x) & 0 \le x < \dfrac{l}{2} \\ w\left(x - \dfrac{l}{2}\right) & \dfrac{l}{2} \le x \le l \end{cases} \tag{4.93}$$

Equation (4.93) took into account the transverse symmetry of the microbridge structure.

The bending-related resonant frequency can simply be calculated by combining the lumped-parameter stiffness of Eq. (4.88) and effective mass of Eq. (4.92), according to the definition.

A similar calculation procedure enables us to find the lumped-parameter torsional resonant frequency of the entire structure, illustrated in Fig. 4.15. The torsional stiffness associated with midpoint 2 is calculated as

$$k_{t,e} = \frac{2}{C_t} \quad (4.94)$$

When the two segments are of constant cross section, the generic stiffness of Eq. (4.94) becomes Eq. (4.28), which indeed defines the stiffness of a constant-cross-section microbridge of length l.

The lumped-parameter torsional mechanical moment of inertia can be expressed as

$$J_{t,e} = \frac{\rho t}{12} \int_0^l f_t^2(x) w(x) [w(x)^2 + t^2] \, dx \quad (4.95)$$

where the variable width $w(x)$ is given in Eq. (4.93). When the two segments are of constant cross section, Eq. (4.95) simplifies to Eq. (4.33) which gives the effective torsional moment of inertia of a constant-cross-section microbridge of length l. The torsional resonant frequency can be calculated by means of Eqs. (4.94) and (4.95).

Example: Analyze the bending and torsional resonant frequencies of the circular corner-filleted microbridge sketched in Fig. 4.16 by applying the two-segment microbridge model.

The microbridge of Fig. 4.16 can be split into two identical units that are mirrored with respect to the symmetry line passing through the structure's

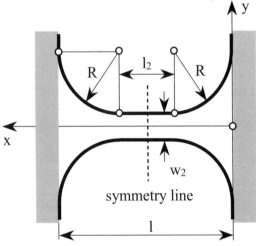

Figure 4.16   Geometry of a doubly filleted microbridge.

one half unit   basic unit # 2

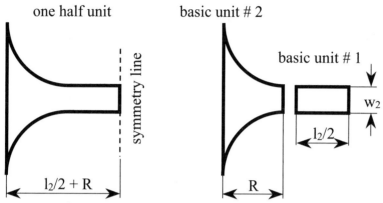

Figure 4.17   Geometry of one unit and of component basic units.

center. Likewise, the left-side unit is formed of two basic units, as shown in Fig. 4.17.

In bending, the three compliances C_l, C_c, and C_r are needed for the half unit, and they can be found by combining the two basic units making up a half unit. This process would mean adding the compliances of the two basic units by expressing the compliances of the basic unit 2 in terms of a reference frame that is placed on the symmetry line [this can be achieved by means of the compliance transform Eqs. (3.2) and (3.10)]. The compliances of both the constant-cross-section basic unit 1 and the basic unit 2 can be calculated by their definition in Eqs. (2.27). For the circularly filleted basic unit of Fig. 4.17 they are

$$
C_l = \frac{3\left[2(4+\pi)R^2 + 4(1+\pi)Rw_2 + \pi w_2^2 \atop -4(2R+w_2)\sqrt{w_2(4R+w_2)}\arctan\sqrt{1+4R/w_2}\right]}{4Et^3}
$$

$$
C_c = \frac{3\left[(2R+w_2)\ln(1+2R/w_2) - 2R\right]}{Et^3}
$$

$$
C_r = \frac{3\left[4(2R+w_2)/\sqrt{w_2(4R+w_2)}\arctan\sqrt{1+4R/w_2}\right]}{Et^3}
$$

(4.96)

The rest of the calculations for the bending stiffness derivation follow the general pattern of the two-segment microbridge model, and the final equation is not presented here because it is too complex.

The lumped-parameter stiffness can be expressed as

$$
m_{b,e} = \frac{\rho t [2R^6 (2.56 l_2^4 + 14.14 l_2^3 R + 29.71 l_2^2 R^2}{(l_2 + 2R)^8}
$$

$$
\frac{+ 28.14 l_2 R^3 + 10.13 R^4) + 0.4(l_2 + 2R)^9 w_2]}{} \tag{4.97}
$$

In torsion, the stiffness associated to the midpoint of the microbridge is

$$
k_{t,e} = \frac{4Gt^3}{3[l_2 w_2 + 2(2R + w_2)\arctan\sqrt{1 + 4R/w_2}}
$$

$$
\frac{}{/\sqrt{w_2(4R + w_2)} - \pi/2]} \tag{4.98}
$$

The torsional mechanical moment of inertia is

$$
J_{t,e} = \frac{\begin{aligned}&\rho t \big\{(23.3 l_2^2 + 80.86 l_2 R + 70.58 R^2)R^6\\ &+ 8[3(5.33 l_2^2 + 17.59 l_2 R + 14.66 R^2)R^5\\ &+ 7(l_2 + 2R)^5 t^2] w_2 + 42(8.76 l_2^2 + 26.51 l_2 R\\ &+ 20.38 R^2)R^4 w_2^2 + 56(l_2 + 2R)^5 w_2^3\\ &+ 14[8.76 l_2^2 + (26.51 l_2 + 20.38 R)R]R^4 t^2\big\}\end{aligned}}{1260(l_2 + 2R)^4} \tag{4.99}
$$

The torsional resonant frequency equation, which is too long and is not explicitly given here, can be found by combining Eqs. (4.98) and (4.99).

Three-segment microbridges. We now consider microbridge formed of three segments of which two that are placed at the fixed ends are identical and mirrored with respect to the midpoint of the microbridge. Their cross section is variable whereas the middle portion of the microbridge which connects the two end segments is of constant cross section, as sketched in Fig. 4.18. The bending and torsional resonant frequencies are now derived by first calculating the lumped-parameter stiffness and inertia fractions corresponding to bending and torsional free vibrations.

Bending resonant frequency. The bending stiffness of the microbridge is calculated as

$$
k_{b,e} = \frac{F_{Cz}}{u_{Cz}} \tag{4.100}
$$

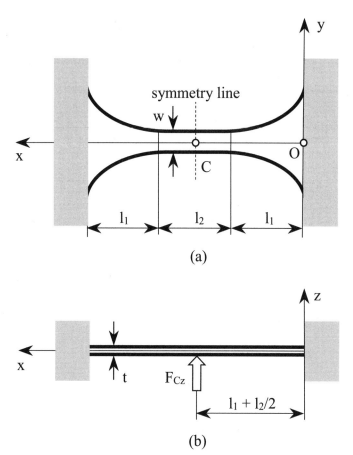

Figure 4.18   Microbridge with three compliant segments: (a) top view; (b) side view.

where F_{Cz} is the force applied at the microbridge center and u_{Cz} is the resulting deflection. Lobontiu and Garcia[18] gave a more detailed presentation of this subject, but in essence, one has to determine the reaction force and moment that are set up under the action of the midpoint force at the right end of the microbridge, for instance, followed by calculation of the deflection at the midpoint. After we carry out these calculations, the bending stiffness is expressed as

$$k_{b,e} = \frac{1}{c + c_2^2 C_r^{'} - 2c_1 c_2 C_c^{'} + c_1^2 C_l^{'} + c_3^2 C_r^{''} - 2c_3 c_4 C_c^{''} + c_4^2 C_l^{'}} \qquad (4.101)$$

where

$$
c = \frac{12}{Ewt^3} \left\{ \frac{(c_2^2 + c_3^2)l_2}{2} - c_1 c_2 \left[\left(l_1 + \frac{l_2}{2} \right)^2 - l_1^2 \right] + \frac{c_1^2 \left[. \left(l_1 + l_2/2 \right)^3 - l_1^3 . \right]}{3} \right.
$$
$$
\left. - c_3 c_4 \left[(l_1 + l_2)^2 - \left(l_1 + \frac{l_2}{2} \right)^2 \right] + \frac{c_4^2 \left[. (l_1 + l_2)^3 - (l_1 + l_2/2)^3 . \right]}{3} \right\}
$$

$$(4.102)$$

$$
c_3 = c_2 - \left(l_1 + \frac{l_2}{2} \right) \qquad c_4 = c_1 - 1 \qquad (4.103)
$$

$$
c_1 = \frac{a_{22}b_1 + a_{12}b_2}{a_{12}a_{21} - a_{11}a_{22}} \qquad c_2 = \frac{a_{21}b_1 + a_{11}b_2}{a_{12}a_{21} - a_{11}a_{22}} \qquad (4.104)
$$

The coefficients defining c_1 and c_2 in Eqs. (4.104) are

$$
a_{11} = a_{22} = C_c' + \frac{6[(l_1 + l_2)^2 - l_1^2]}{Ewt^3} + C_c'' \qquad (4.105)
$$

$$
a_{12} = C_r' + \frac{12l_2}{Ewt^3} + C_r'' \qquad (4.106)
$$

$$
a_{21} = C_l' + \frac{4[(l_1 + l_2)^3 - l_1^3]}{Ewt^3} + C_l'' \qquad (4.107)
$$

$$
b_1 = \frac{6[l_2(l_1 + l_2/2) - (l_1 + l_2)^2 + (l_1 + l_2/2)^2]}{Ewt^3} + \left(l_1 + \frac{l_2}{2} \right) C_r'' - C_c'' \quad (4.108)
$$

$$
b_2 = \frac{2\left\{ 2[(l_1 + l_2)^3 - (l_1 + l_2/2)^3] \atop -3(l_1 + l_2/2)[(l_1 + l_2)^2 - (l_1 + l_2/2)^2] \right\}}{Ewt^3} - \left(l_1 + \frac{l_2}{2} \right) C_c'' + C_l'' \qquad (4.109)
$$

A check can be performed on this model by considering that all segments are identical with constant rectangular cross section (with w and t as cross-sectional dimensions) and length $l/3$. Equations (4.101) through (4.109) yield the stiffness corresponding to a fixed-fixed homogeneous beam of length l and cross section defined by w and t.

The lumped mass which is dynamically equivalent to the distributed mass microbridge undergoing free bending vibrations, and which needs to be placed at the midspan, is calculated by equating the kinetic energy of the equivalent (effective) mass to the kinetic energy of the vibrating microcantilever such that

$$
m_{b,e} = \rho t \left[
\begin{array}{c}
\displaystyle\int_0^{l_1} f_b(x)^2 w(x)\,dx + w \int_{l_1}^{l_1+l_2} f_b(x)^2\,dx \\[2em]
+ \displaystyle\int_{l_1+l_2}^{2l_1+l_2} f_b(x)^2 w(x)\,dx
\end{array}
\right]
\tag{4.110}
$$

Due to the transverse symmetry of the microbridge shown in Fig. 4.18, Eq. (4.110) can also be expressed as

$$
m_{b,e} = \rho t \left[2\int_0^{l_1} f_b(x)^2 w(x)\,dx + w \int_{l_1}^{l_1+l_2} f_b(x)^2\,dx \right]
\tag{4.111}
$$

A check is performed again to see whether the generic Eq. (4.111) reduces to the known mass equation in the case where the two end segments of the microbridge are identical to the middle one. Indeed, when $l_1 = l_2 = l/3$ and $w(x) = w$, Eq. (4.111) reduces to Eq. (4.12), which gives the effective mass of a constant rectangular cross-section microbridge of length l and cross section defined by w and t.

By combining Eqs. (4.101) with Eq. (4.111), the resonant frequency corresponding to bending vibrations of the microbridge sketched in Fig. 4.18 is

$$
\omega_{b,e} = \frac{1}{\sqrt{\rho\, t[2\displaystyle\int_0^{l_1} f_b(x)^2 w(x)\,dx + w\int_{l_1}^{l_1+l_2} f_b(x)^2\,dx]}}
$$
$$
\times \sqrt{c + c_2^2 C_r' - 2c_1 c_2 C_c' + c_1^2 C_l' + c_3^2 C_r'' - 2c_3 c_4 C_c'' + c_4^2 C_l''}
\tag{4.112}
$$

For a microbridge formed of three identical segments ($l_1 = l_2 = l/3$) of constant rectangular cross section, the generic Eq. (4.112) reduces to Eq. (4.13), which defines the bending resonant frequency of a constant-cross-section microbridge of length l.

Torsional resonant frequency. In torsion, the lumped-parameter stiffness being associated with the midspan of the microbridge pictured in Fig. 4.18 is determined by first calculating the endpoint reaction moment produced by a torsional moment (about the x axis) applied at the midpoint, followed by calculation of the resulting angular deformation at the same midpoint, in the form:

$$k_{t,e} = \frac{M_{Cx}}{\theta_{Cx}} \tag{4.113}$$

After we perform the required calculations, the torsional stiffness of Eq. (4.113) is expressed as

$$k_{t,e} = \frac{4GI_{t2}}{l_2 + 2C_t GI_{t2}} \tag{4.114}$$

where C_t is the torsional stiffness of one of the end segments and I_{t2} is the torsional moment of inertia of the middle segment. When $w(x) = w$ and $l_1 = l_2 = l/3$, Eq. (4.114) simplifies to Eq. (4.28) which gives the torsional stiffness of a constant-cross-section microbridge of length l.

The lumped-parameter torsional mechanical moment of inertia at the midpoint, which is dynamically equivalent to the rotary inertia of the distributed-parameter microbridge undergoing torsional vibrations, is calculated by applying again Rayleigh's principle and therefore by equating the kinetic energy of the equivalent system to the kinetic energy of the real one. The effective mechanical moment of inertia is

$$J_{t,e} = \frac{\rho t}{12} \left\{ 2 \int_0^{l_1} f_t(x)^2 w(x)[w(x)^2 + t^2] \, dx \right.$$

$$\left. + w_2(w_2^2 + t^2) \int_{l_1}^{l_1 + l_2} f_t(x)^2 dx \right\} \tag{4.115}$$

where the torsional distribution function is

$$f_t(x) = \frac{4x(2l_1 + l_2 - x)}{(2l_1 + l_2)^2} \tag{4.116}$$

In the case where the two end segments are identical to the middle one, and therefore have constant cross section and are of length $l/3$, Eqs. (4.115) and (4.116) reduce to Eq. (4.33), which expresses the effective moment of inertia corresponding to free torsional vibrations of a constant-cross-section microbridge of length l.

The torsional resonant frequency of the microbridge is found by combining Eqs. (4.114), (4.115), and (4.116) according to the definition, namely,

$$\omega_{t,e} = \cfrac{6.93\sqrt{GI_{t2}/[\rho t(l_2 + 2C_t GI_{t2})]}}{\sqrt{2\displaystyle\int_0^{l_1} f_t(x)^2 w(x)[w(x)^2 + t^2]\,dx + w_2(w_2^2 + t^2)\displaystyle\int_{l_1}^{l_1+l_2} f_t(x)^2\,dx}}$$

(4.117)

Paddle microbridges. The geometry of a constant-thickness paddle microbridge is shown in Fig. 4.19 in top view. Its configuration is similar to that of a paddle microcantilever and consists of a middle section of width w_2 and two identical end/root portions of width w_2. This design is a particular illustration of the generic microbridge design just analyzed, and the corresponding model is applied to this paddle design to determine the relevant resonant frequencies. Before we apply the generic model, a simpler approach that matches the relatively uncomplicated geometry is taken, by directly using Castigliano's displacement theorem (which yields the relevant stiffnesses) and Rayleigh's principle (which provides the relevant effective inertia fractions). This approach only analyzes one-half of the microbridge because of its transverse symmetry.

Direct approach. Due to the paddle microbridges transverse symmetry, it is sufficient to study only one-half of its structure to determine

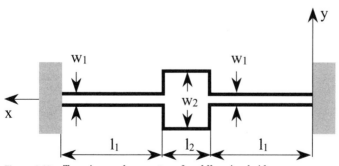

Figure 4.19   Top view and geometry of paddle microbridge.

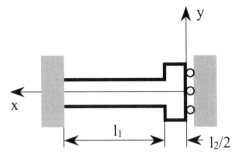

Figure 4.20   Half-model of paddle microbridge.

the torsional and bending resonant frequencies, by directly using Castigliano's displacement theorem for the lumped-parameter stiffness calculation and Rayleigh's principle for effective inertia fraction determination. Figure 4.20 pictures the half-length paddle microbridge and also indicates the guided boundary condition at the midspan which has to be utilized in bending calculations.

In torsion, the stiffness of the half-microbridge is

$$k_{t,e} = \frac{2Gt^3 w_1 w_2}{3(w_1 l_2 + 2w_2 l_1)} \qquad (4.118)$$

Equation (4.118) simplifies to Eq. (4.24) — giving the torsional stiffness of a homogeneous, constant-cross-section half-microcantilever — when $w_2 = w_1$, $l_1 = l/4$ and $l_2 = l/2$ (such that $l_1 + l_2/2 = l/2$).

The equivalent mechanical moment of inertia which is dynamically equivalent to the distributed-parameter half-microbridge undergoing free torsional vibrations is calculated by applying Rayleigh's principle, which has been detailed thus far. It is worth noting that the distribution function connecting the rotation/deformation angle at an abscissa x measured from the guided end in Fig. 4.20 to the maximum rotation/ deformation angle is

$$f_t(x) = 1 - \frac{x}{l_1 + l_2/2} \qquad (4.119)$$

and therefore of the form corresponding to a fixed-free microbar (fixed-free beam). The effective moment of inertia is calculated as

$$J_{t,e} = \frac{\rho t}{12} \left[w_2(w_2^2 + t^2) \int_0^{l_2/2} f_t(x)^2 \, dx \right.$$

$$\left. + w_1(w_1^2 + t^2) \int_{l_2/t}^{l_1 - l_2/2} f(x)^2 \, dx \right] \tag{4.120}$$

After we perform the calculations involved in Eqs. (4.119) and (4.120), the moment of inertia becomes

$$J_{t,e} = \frac{\rho t [8l_1^3 w_1(w_1^2 + t^2) + l_2 w_2(12l_1^2 \atop + 6l_1 l_2 + l_2^2)(w_2^2 + t^2)]}{72(2l_1 + l_2)^2} \tag{4.121}$$

When $w_2 = w_1$, $l_1 = l/4$, and $l_2 = l/2$, Eq. (4.121) simplifies to Eq. (4.26), which gives the effective torsional mechanical moment of inertia for a constant-cross-section bar of length $l/2$. The lumped-parameter torsional resonant frequency is calculated by combining the lumped-parameter stiffness of Eq. (4.118) and the lumped-parameter inertia of Eq. (4.121) as

$$\omega_{t,e} = \frac{6.928(2l_1 + l_2)t\sqrt{Gw_1 w_2 / [\rho(w_1 l_2 + 2w_2 l_1)]}}{\sqrt{8l_1^3 w_1(w_1^2 + t^2) + l_2 w_2(12l_1^2 + 6l_1 l_2 + l_2^2)(w_2^2 + t^2)}} \tag{4.122}$$

In bending, the stiffness of the half microbridge is found by applying a force at the guided end in Fig. 4.20, perpendicularly on the beam's axis, and by determining the corresponding deflection. The lumped stiffness is

$$k_{b,e} = \frac{8Et^3 w_1 w_2(w_1 l_2 + 2w_2 l_1)}{w_1^2 l_2^4 + 8l_1 l_2(4l_1^2 + 3l_1 l_2 + l_2^2)w_1 w_2 + 16w_2^2 l_1^4} \tag{4.123}$$

Equation (4.123) simplifies to Eq. (4.1), which calculates the bending stiffness of a constant-cross-section half-microcantilever (of fixed-guided boundary conditions) in the case where $w_2 = w_1$, $l_1 = l/4$, and $l_2 = l/2$.

The lumped mass which needs to be placed at the guided end of the half-microbridge in Fig. 4.20 is calculated again by means of Rayleigh's principle. The distribution function which expresses the deflection at a generic point x on the half-microbridge in terms of the maximum tip deflection is, in this case,

$$f_b(x) = \frac{(l_1 + l_2/2 - 2x)^2 (l_1 + l_2/2 + 4x)}{(l_1 + l_2/2)^3} \tag{4.124}$$

The lumped mass which is dynamically equivalent to the distributed inertia of the half-microbridge undergoing free bending vibrations is calculated as

$$m_{b,e} = \rho t \left[w_2 \int_0^{l_2/2} f_b(x)^2 \, dx + w_1 \int_{l_2/2}^{l_1 + l_2/2} f_b(x)^2 \, dx \right] \tag{4.125}$$

and its final equation is

$$m_{b,e} = \frac{\rho t [32 l_1^5 (52 l_1^2 + 112 l_1 l_2 + 63 l_2^2) w_1 + l_2 (2240 l_1^6 \\ + 6720 l_1^5 l_2 + 7280 l_1^4 l_2^2 + 3640 l_1^3 l_2^3 + 1092 l_1^2 l_2^4 \\ + 182 l_1 l_2^5 + 13 l_2^6) w_2]}{70(2l_1 + l_2)^6} \tag{4.126}$$

This equation, again, reduces to Eq. (4.4)—expressing the lumped mass for a constant-cross-section half-length microbridge—when $w_2 = w_1$, $l_1 = l/4$ and $l_2 = l/2$. The lumped-parameter resonant frequency corresponding to free bending vibrations is calculated with the aid of Eqs. (4.123) and (4.126) as

$$\omega_b = \frac{23.664(2l_1 + l_2)^3 t \sqrt{E w_1 w_2 (w_1 l_2 + w_2 l_1) / \left\{ \rho [w_1^2 l_2^4 + 8 l_1 l_2 (4 l_1^2 \\ + 3 l_1 l_2 + l_2^2) w_1 w_2 + 16 w_2^2 l_1^4] \right\}}}{\sqrt{\begin{array}{l} 32 l_1^5 (52 l_1^2 + 112 l_1 l_2 + 63 l_2^2) w_1 \\ + l_2 (2240 l_1^6 + 6720 l_1^5 l_2 + 7280 l_1^4 l_2^2 \\ + 3640 l_1^3 l_2^3 + 1092 l_1^2 l_2^4 + 182 l_1 l_2^5 + 13 l_2^6) w_2 \end{array}}} \tag{4.127}$$

Generic-model approach. The torsion and bending resonant frequencies can also be determined for the paddle microbridge of Fig. 4.19 by applying the generic model developed in this section for a three-segment microbridge.

The torsional stiffness associated with the midpoint of the microbridge is

$$k_{t,e} = \frac{4Gt^3 w_1 w_2}{3(w_1 l_2 + 2w_2 l_1)} \tag{4.128}$$

It can be seen that the stiffness expressed in Eq. (4.128), which corresponds to the full-length paddle microbridge, is twice the torsional stiffness of the half-microbridge [Eq. (4.110)].

The effective torsional mechanical moment of inertia is

$$J_{t,e} = \frac{2\rho t [2l_1^3 w_1 (16 l_1^2 + 25 l_1 l_2 + 10 l_2^2)(w_1^2 + t^2)}{45(2l_1 + l_2)^4} \tag{4.129}$$

$$+ l_2 w_2 (w_2^2 + t^2)(30 l_1^4 + 60 l_1^3 l_2 + 40 l_1^2 l_2^2 + 10 l_1 l_2^3 + l_2^4)]$$

When $l_1 = l_2 = l/3$ and $w_1 = w_2$, Eq. (4.129) reduces to Eq. (4.33), which provides the effective mechanical moment of inertia of a constant-cross-section microbridge of length l. The resulting torsional resonant frequency is, by way of the stiffness Eq. (4.128) and mechanical moment of inertia Eq. (4.129),

$$\omega_{t,e} = \frac{5.48t(2l_1 + l_2)^2 \sqrt{Gw_1 w_2 / \rho(w_1 l_2 + 2w_2 l_1)}}{\sqrt{\begin{array}{c} 2l_1^3 w_1 (16 l_1^2 + 25 l_1 l_2 + 10 l_2^2)(w_1^2 + t^2) \\ + l_2 w_2 (w_2^2 + t^2)(30 l_1^4 + 60 l_1^3 l_2 + 40 l_1^2 l_2^2 + 10 l_1 l_2^3 + l_2^4) \end{array}}} \tag{4.130}$$

Obviously, for the same limit conditions as above, namely $l_1 = l_2 = l/3$ and $w_1 = w_2$, Eq. (4.130) reduces to Eq. (4.34), which yields the torsional resonant frequency of a constant-cross-section microbridge of length l.

The lumped-parameter bending stiffness, which is associated with the midpoint of the paddle microbridge, is

$$k_{b,e} = \frac{16Et^3 w_1 w_2 (w_1 l_2 + 2w_2 l_1)}{w_1^2 l_2^4 + 8 l_1 l_2 (4 l_1^2 + 3 l_1 l_2 + l_2^2) w_1 w_2 + 16 w_2^2 l_1^4} \tag{4.131}$$

and this is twice that given in Eq. (4.123) for a half-length paddle microbridge.

The bending-related effective mass is

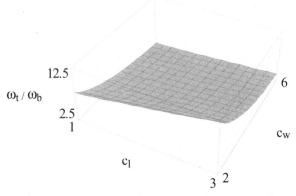

Figure 4.21   Torsion-to-bending resonant frequency ratio in terms of length and width parameters.

$$128\rho t[512l_1^9 w_1 + l_2^5 w_2(2016l_1^4 + 672l_1^3 l_2 + 144l_1^2 l_2^2$$

$$+18l_1 l_2^3 + l_2^4) + 252l_1^5 l_2^4(w_1 + 15w_2)$$

$$+168l_1^6 l_2^3(7w_1 + 25w_2) + 72l_1^7 l_2^2(29w_1 + 35w_2) \qquad (4.132)$$

$$m_{b,e} = \frac{+18l_1^8 l_2(93w_1 + 35w_2)]}{315(2l_1 + l_2)^8}$$

For $l_1 = l_2 = l/3$ and $w_1 = w_2$, Eq. (4.132) simplifies to Eq. (4.12), which defines the bending effective mass of a constant-cross-section microbridge of length l. The bending resonant frequency is calculated by Eqs. (4.131) and (4.132) and is too complex to be included here. It can be shown, however, that when $l_1 = l_2 = l/3$ and $w_1 = w_2$, this bending frequency reduces to Eq. (4.13), which defines the resonant frequency of a constant-cross-section microbridge of length l.

Example: Compare the torsional and bending resonant frequencies of the paddle microbridge sketched in Fig. 4.20. Known are the thickness $t = 1$ µm, width of root segment $w_1 = 10$ µm, and length of the same segment $l_1 = 100$ µm. By using the nondimensional parameters $c_l = l_2/l_1$ and $c_w = w_2/w_1$, the ratio of the torsional resonant frequency [Eq. (4.122)] to the bending one [Eq. (4.127)] is pictured in Fig. 4.21, which shows that the torsional frequency is higher than the bending frequency for the chosen parameter ranges.

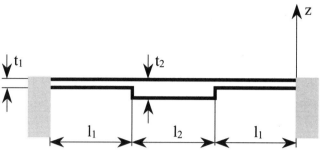

Figure 4.22   Paddle microbridge with step variable-thickness.

Another paddle-type microbridge configuration is shown in Fig. 4.22, a design which has its middle portion thicker than the two adjoining end parts.

The generic model of a microbridge consisting of a central constant rectangular cross-section portion and two end identical and mirrored portions (which can be of variable cross section) is utilized for the configuration shown in Fig. 4.22, with the mention that the two end segments are also of constant cross section.

By using the nondimensional parameters c_l and c_t defined as

$$l_2 = c_l l_1 \qquad t_2 = c_t t_1 \qquad (4.133)$$

the bending stiffness of a long microbridge (according to the Euler-Bernoulli model, in which shearing effects are neglected) associated with the midspan can be expressed as

$$k_{b,e} = \frac{16 E c_t^3 (c_l + 2c_t^3) w t_1^3}{\{c_l^4 + 8[4 + c_l(3 + c_l)]c_l c_t^3 + 16 c_t^6\} l_1^3} \qquad (4.134)$$

Notice that for $c_l \to 1$ (which means $l_1 = l_2$) and $c_t \to 1$ (which means $t_1 = t_2$), together with $l_1 = l/3$, Eq. (4.134) reduces to Eq. (4.7), which gives the stiffness of a constant-cross-section microbridge of length l.

In the case where the microbridge is relatively short, shearing effects need to be accounted for according to the Timoshenko model, and the linear direct bending stiffnesses of both the central and the end portions have to be calculated accordingly. As a consequence, the resulting stiffness is

$$k_{b,e}^{sh} = \frac{16EGc_t^3(c_l + 2c_t^3)wt_1^3}{G\{c_l^4 + 8c_l[4 + c_l(3 + c_l)]c_t^3 + 16c_t^6\}l_1^3} \tag{4.135}$$
$$+8\kappa Ec_t^3(c_l + 2c_t^3)l_1t_1^2$$

The effective mass that corresponds to the out-of-the-plane free bending vibrations is

$$m_{b,e} = \frac{\begin{array}{l} 0.4m\{512 + 6c_l[279 + 2c_l(174 + 7c_l(14 + 3c_l))] \\ +c_l[630 + c_l(2520 + c_l(4200 + c_l(3780 + c_l(2016 \\ +c_l(672 + c_l(144 + c_l(18 + c_l)))))))]\} \end{array}}{(2 + c_l)^8} \tag{4.136}$$

The effective mass of Eq. (4.136) reduces to the mass of Eq. (4.12) when $c_l \to 1, c_t \to 1$, and $l_1 = l/3$, which proves its validity. The resonant bending frequency is the square root of the ratio of the bending stiffness given in either Eq. (4.133) or Eq. (4.134) to the effective mass of Eq. (4.136).

The torsional stiffness at the microbridge midpoint is, according to the generic algorithm,

$$k_{t,e} = \frac{4c_t^3t_1^3wG}{3l_1(c_l + 2c_t^3)} \tag{4.137}$$

When $c_l \to 1$, $c_t \to 1$, and $l_1 = l/3$, Eq. (4.137) simplifies to Eq. (4.28) which provides the torsional stiffness of a constant-cross-section microbridge of length l.

The lumped-parameter (effective) mechanical moment of inertia is, by way of utilizing the same generic model,

$$J_{t,e} = \frac{\begin{array}{l} 0.04\{[32 + 10c_l(5 + 2c_l)](t_1^2 + w^2) + [c_l(30 + c_l(60 + c_l(40 \\ +c_l(10 + c_l))))]c_t(c_t^2t_1^2 + w^2)\} \end{array}}{(2 + c_l)^4} \tag{4.138}$$

Again, for $c_l \to 1$, $c_t \to 1$, and $l_1 = l/3$, Eq. (4.138) reduces to Eq. (4.33) which corresponds to the effective inertia of a constant-cross-section microbridge, and this proves the validity of Eq. (4.138).

The resonant torsional frequency is found by combining Eqs. (4.137) and (4.138) as the square root of the stiffness-to-inertia ratio, namely,

$$\omega_{t,e} = \frac{5.48(2 + c_l)^2 c_t t_1 \sqrt{Gk_t / [\rho(c_l + 2c_t^3)]} / l_1}{\sqrt{\begin{array}{l}\{[32 + 10c_l(5 + 2c_l)](t_1^2 + w^2) + [c_l(30 + c_l(60 \\ +c_l(40 + c_l(10 + c_l))))]c_t(c_t^2 t_1^2 + w^2)\}/(2 + c_l)^4\end{array}}} \tag{4.139}$$

Doubly trapezoid microbridges. The doubly trapezoid microbridge of Fig. 4.23 consists of two trapezoid end portions adjoining a constant-width middle portion. The maximum width of this design is w_1, whereas the minimum width (also the width of the middle segment) is w_2. The lumped-parameter stiffness and inertia properties corresponding to bending and torsion, respectively, are determined according to the generic algorithm detailed previously.

The bending stiffness which is associated with the microbridge midpoint is

$$k_{b,e} = \frac{16Et^3 w_1 c_w (c_w - 1)^2 [c_l(c_w - 1) + 2c_w \ln c_w]}{l_1^3 A} \tag{4.140}$$

where

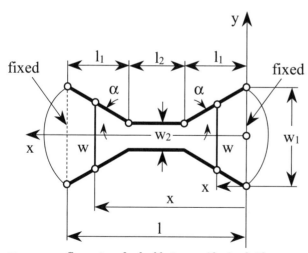

Figure 4.23   Geometry of a doubly trapezoid microbridge.

$$A = (c_w - 1)[c_l^4(c_w - 1)^2 - 48c_l^2 c_w(c_w - 1)$$

$$- 192c_w^2 - 48c_l c_w(3c_w - 1)] + 8c_w[c_l^3(c_w - 1)^2 \qquad (4.141)$$

$$+ 6c_l^2 c_w(c_w - 1) + 12c_l c_w^2 + 12c_w(c_w + 1)]\ln c_w$$

In Eqs. (4.140) and (4.141), the following notations have been used:

$$w_2 = c_w w_1 \qquad l_2 = c_l l_1 \qquad (4.142)$$

When $w_2 = w_1$ and $l_1 = l_2 = l/3$, Eq. (4.140) reduces to Eq. (4.47), which gives the bending stiffness of a constant-cross-section microbridge of length l.

By applying the inertia derivation of the generic microbridge model, the lumped-parameter mass which needs to be placed at the midspan and which is dynamically equivalent to the distributed-parameter microbridge of Fig. 4.23 is

$$m_{b,e} = \frac{128 \rho t w_1 l_1 B}{315(2 + c_l)^8} \qquad (4.143)$$

where

$$B = 386 + 126c_w + c_l[1930 + 374c_w + c_l(9(465 + 47c_w)$$

$$+ c_l(24(215 + 9c_w) + c_l(c_l(2016 + c_l(672 + c_l(144 \qquad (4.144)$$

$$+ c_l(18 + c_l)))) + 42(95 + c_w))))]$$

Again, Eq. (4.143) reduces to Eq. (4.12)—giving the lumped-parameter mass of a constant-cross-section microbridge—when $w_2 = w_1$ and $l_1 = l_2 = l/3$.

The resonant frequency corresponding to free bending vibrations is found by combining Eqs. (4.140) and (4.143) according to the known equation, namely,

$$\omega_{b,e} = \frac{6.275(2 + c_l)^4(c_w - 1)t}{l_1^2} \sqrt{\frac{Ec_w[c_l(c_w - 1) + 2c_w \ln c_w]}{\rho AB}} \qquad (4.145)$$

In torsion, the lumped-parameter stiffness corresponding to the midpoint of the microbridge sketched in Fig. 4.22 is

$$k_{t,e} = \frac{4Gc_w(c_w - 1)w_1 t^3}{3l_1[c_l(c_w - 1) + 2c_w \ln c_w]} \qquad (4.146)$$

Equation (4.146) simplifies to Eq. (4.28) which provides the torsional stiffness of a constant-cross-section microbridge when $w_2 = w_1$ and $l_1 = l_2 = l/3$.
The torsional mechanical moment of inertia is

$$J_{t,e} = \frac{\rho t D}{315(c_l + 2)^4} \qquad (4.147)$$

where

$$D_1 = 14\{22 + 36c_l + 15c_l^2 + (c_l + 1)[10 + c_l(c_l + 2)(17$$

$$+ c_l(c_l + 7))]c_w\}l_1 t^2 w_1 + \{5[37 + 4c_l(16 + 7c_l)]$$

$$+ 3[47 + 4c_l(18 + 7c_l)]c_w + 3[29 + 2c_l(20 + 7c_l)]c_w^2$$

$$+ [35 + 2c_l(232 + 7c_l(61 + c_l(40 + c_l(10 + c_l))))]c_w^3\}l_1 w_1^3 \qquad (4.148)$$

Again, the mechanical moment of inertia of Eq. (4.147) reduces to the one of a constant-cross-section microbridge of length l—when $w_2 = w_1$ and $l_1 = l_2 = l/3$.
The lumped-parameter resonant frequency corresponding to free torsional vibrations is calculated by means of Eqs. (4.146) and (4.147) as

$$\omega_{t,e} = \frac{20.5t(2 + c_l)^2}{l_1} \sqrt{\frac{Gc_w(c_w - 1)}{\rho D[c_l(c_w - 1) + 2c_w \ln c_w]}} \qquad (4.149)$$

Figure 4.24 shows the design derived from Fig. 4.23 by eliminating the constant rectangular middle portion.
The lumped-parameter stiffness, inertia, and resonant frequencies corresponding to bending and torsion for this design are simply obtained by taking into consideration that $l_2 = 0$ and therefore $c_l = 0$—according to the definition in Eq. (4.142)—in the equations defining similar lumped-parameter properties for the more generic microbridge of Fig. 4.23. The bending stiffness is therefore

$$k_{b,e} = \frac{Et^3 w_1(c_w - 1)^2 \ln c_w}{3l_1^3[(c_w + 1)\ln c_w - 2(c_w - 1)]} \qquad (4.150)$$

The effective mass which is dynamically equivalent to the bending-vibrating distributed-parameter microbridge is

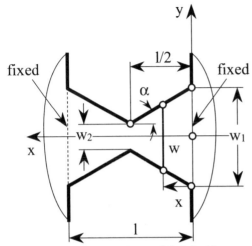

Figure 4.24   Geometry of a simplified doubly trapezoid microbridge.

$$m_{b,e} = \frac{(63 + 193c_w)\rho t w_2 l_1}{315} \qquad (4.151)$$

and the resulting bending-produced resonant frequency is

$$\omega_{b,e} = \frac{10.247(1-c_w)t}{l_1^2} \times \sqrt{\frac{E \ln c_w}{\rho(63c_w + 193)[(c_w + 1)\ln c_w - 2(c_w-1)]}} \qquad (4.152)$$

The torsional stiffness is

$$k_{t,e} = \frac{2G(c_w - 1)t^3 w_2}{3l_1 \ln c_w} \qquad (4.153)$$

The lumped-parameter inertia fraction which is dynamically equivalent to the free torsional vibrations of the distributed-parameter microbridge is

$$J_{t,e} = \frac{\left\{28(11c_w + 5)t^2 + [35 + c_w(87 + c_w(141 + 185c_w))]\, w_1^2\right\}\rho t w_2 l_1}{5040} \qquad (4.154)$$

The torsional resonant frequency is

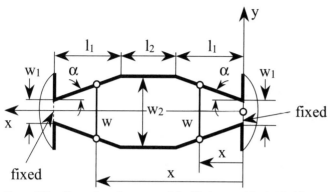

Figure 4.25 Geometry of a reversed doubly trapezoid microbridge.

$$\omega_{t,e} = \frac{\left(57.965t / l_1\right)\sqrt{G(c_w - 1)/\rho}}{\sqrt{28(11c_w + 5)t^2 + [35 + c_w(87 + c_w(141 + 185c_w))]w_1^2}} \quad (4.155)$$

In Eqs. (4.150) through (4.155), l_1 is actually equal to $l/2$, as shown in Fig. 4.24.

The reversed doubly trapezoid microbridge of Fig. 4.25 is similar to the design of Fig. 4.23, having its two trapezoid end portions that are adjoining the constant-width middle portion of reversed inclination. The maximum width of this design is w_2 (it is also the width of the middle segment), whereas the minimum width is w_1.

The lumped-parameter bending stiffness corresponding to the midpoint of the microbridge pictured in Fig. 4.25 is

$$k_{b,e} = \frac{16Et^3 w_1 c_w (c_w - 1)^2 [c_l(c_w - 1) + 2c_w \ln c_w]}{l_1^3 A} \quad (4.156)$$

where

$$A = (c_w - 1)[c_l^4(c_w - 1)^2 - 48c_l^2 c_w(c_w - 1) - 192c_w^2$$

$$- 48c_l c_w(3c_w - 1)] + 8c_w[c_l^3(c_w - 1)^2 \quad (4.157)$$

$$+ 6c_l^2 c_w(c_w - 1) + 12c_l c_w^2 + 12c_w(c_w + 1)\ln c_w]$$

The stiffness expressed by Eqs. (4.156) and (4.157) reduces to Eq. (4.7), which yields the stiffness of a constant-cross-section microbridge of length l, when $w_2 = w_1$ and $l_1 = l_2 = l/3$. Similarly, the lumped-parameter mass which is equivalent to the distributed inertia of the microbridge undergoing free bending vibrations is expressed by Eq. (4.143) where

$$
\begin{aligned}
B = c_w \big\{ & 386 + 1300c_l + 1665c_l^2 + 960c_l^3 \\
& + 210c_l^4 + (c_l + 1)\big\{126 + c_l(c_l + 2)[439 \\
& + c_l(813 + c_l(769 + c_l(351 + c_l(97 + c_l(15 + c_l))))))]\big\}\big\}
\end{aligned} \tag{4.158}
$$

Again, this equation simplifies to Eq. (4.12), expressing the effective mass of a constant-cross-section microbridge of length l, when $w_2 = w_1$ and $l_1 = l_2 = l/3$. The resonant frequency describing the free bending vibrations is

$$
\omega_{b,e} = \frac{6.275(2 + c_l)^4(c_w - 1)t}{l_1^2} \times \sqrt{\frac{Ec_w[c_l(c_w - 1) + 2c_w \ln c_w]}{\rho AB}} \tag{4.159}
$$

The torsional stiffness at the microbridge midpoint is identical to that of the microbridge shown in Fig. 4.23.

$$
k_{t,e} = \frac{4Gc_w(c_w - 1)w_1t^3}{3l_1[c_l(c_w - 1) + 2c_w \ln c_w]} \tag{4.160}
$$

It reduces to Eq. (4.28), which expresses the torsional stiffness of a constant-cross-section microbridge of length l, when $w_2 = w_1$ and $l_1 = l_2 = l/3$.

The mechanical moment of inertia which is placed at the microbridge midpoint and is dynamically equivalent to the distributed-parameter inertia of the microbridge undergoing free torsional vibrations is expressed again by means of Eq. (4.147) where D_2 goes instead of D_1:

$$
\begin{aligned}
D_2 = 14\big\{ & 22 + 36c_l + 15c_l^2 + (c_l + 1)\big\{10 + c_l(c_l + 2)[17 \\
& + c_l(c_l + 7)]\big\}c_w\big\}l_1w_1t^2 + 4\big\{185 + 20c_l(7c_l + 16) \\
& + 141c_w + 2c_l[318 + 7c_l(66 + c_l(40 + c_l(10 + c_l)))]c_w \\
& + 3[29 + 2c_l(7c_l + 20)]c_w^2 + [35 + 2c_l(7c_l + 22)]c_w^3\big\}l_1w_1^3\big\}
\end{aligned} \tag{4.161}
$$

When $w_2 = w_1$ and $l_1 = l_2 = l/3$, this inertia fraction simplifies to that of Eq. (4.33) which defines a constant-cross-section microbridge of length l. The torsional resonant frequency is therefore again:

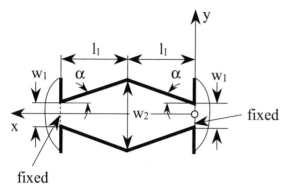

Figure 4.26   Geometry of a simplified reversed doubly trapezoid microbridge.

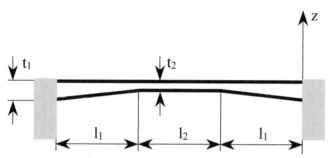

Figure 4.27   Paddle microbridge with linearly variable thickness over the end segments.

$$\omega_{t,e} = \frac{20.5t(2 + c_l)^2}{l_1} \sqrt{\frac{Gc_w(c_w - 1)}{\rho D_2[c_l(c_w - 1) + 2c_w \ln c_w]}} \qquad (4.162)$$

A simplified version of the reversed doubly trapezoid microbridge which has no middle segment is shown in Fig. 4.26. Again, as was the case with the regular double trapezoid configuration, its lumped-parameter properties can be derived from those characterizing the more generic reversed double trapezoid design of Fig. 4.25 by taking $c_l = 0$ (which is equivalent to $l_2 = 0$) in the corresponding equations. The bending stiffness is given by Eq. (4.150), whereas the effective mass is given by Eq. (4.151), which proves that the two simplified designs — the regular and reversed ones — are identical in terms of the free bending response. Similar results are obtained for the free torsional response, and therefore the two above-mentioned designs are again identical in terms of their resonant behavior in torsion.

Another microbridge configuration that contains trapezoid portions is shown in Fig. 4.27. The middle section has constant thickness t_2, whereas the two adjacent end sections have variable thicknesses.

The bending stiffness is

$$k_{b,e} = \frac{16E(c_l + c_t + c_t^2)(c_t - 1)^3 c_t^3 w t_1^3}{l_1^3[(c_t - 1)(c_l^4(c_t - 1)^2 + 24c_l^2(c_t - 1)^2 c_t - 192c_t^4}$$
$$+ 4c_l^3(c_t - 1)^2(c_t + 1)c_t - 48c_l c_t(3c_t - 1))$$
$$+ 96k_t^3(c_l + c_t + c_t^2)\ln c_t]$$

(4.163)

with c_l and c_t being defined in Eqs. (4.133). This equation reduces to Eq. (4.7) when $c_l \to 1$ (or $l_2 = l_1$) and $c_t \to 1$ (or $t_2 = t_1$) and $l_1 = l_2 = l/3$, which proves that the microbridge of Fig. 4.27 transforms to a constant-cross-section configuration when the above-mentioned conditions are satisfied. In the case where the three portions compounding into the microbridge of Fig. 4.20 are relatively short and shearing effects need to be accounted for, the bending stiffness becomes

$$k_{b,e}^{\text{sh}} = \frac{16EGc_t^3(c_l + c_t + c_t^2)(1 - c_t)^3 w t_1^3}{l_1[G(1 - c_t)(c_l^4(c_t - 1)^2 + 24c_l^2 c_t(c_t - 1)^2}$$
$$-192c_t^4 + 4c_l^3 c_t(c_t - 1)^2(c_t + 1) - 48c_l c_t(3c_t - 1))l_1^2$$
$$-8c_t^3(c_l + c_t + c_t^2)(12Gl_1^2 + \kappa E(c_t - 1)^2 t_1^2)\ln c_t]$$

(4.164)

The effective mass in bending is

$$m_{b,e} = \frac{128\{126 + 374c_l + 423c_l^2 + 216c_l^3 + 42c_l^4}{315(2 + c_l)^8}$$
$$+(1 + c_l)[386 + c_l(1544 + c_l(2641 + c_l(2519$$
$$+c_l(1417 + c_l(545 + c_l(127 + c_l(17 + c_l)))))))]c_t\}m_1$$

(4.165)

In Eq. (4.165), m_1 is the mass of the middle constant-cross-section segment. Equation (4.165) transforms to the mass given in Eq. (4.12), $c_l \to 1$ (or $l_2 = l_1$) and $c_t \to 1$ (or $t_2 = t_1$) and $l_1 = l_2 = l/3$, which proves the validity of Eq. (4.165).

The bending resonant frequency can simply be found by combining the lumped-parameter stiffness of Eq. (4.162) for long members or Eq. (4.164) for short members with the mass of Eq. (4.165).

The lumped-parameter stiffness is

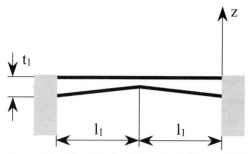

Figure 4.28   Simplified paddle microbridge with linearly variable thickness over the end segments.

$$k_{t,e} = \frac{4Gc_t^3 wt_1^3}{3(c_l + c_t + c_t^2)l_1}$$
(4.166)

When $c_l \to 1$ (or $l_2 = l_1$) and $c_t \to 1$ (or $t_2 = t_1$) and $l_1 = l_2 = l/3$, Eq. (4.166) transforms into Eq. (4.28), which corresponds to a constant-cross-section microbridge of length l.

The effective mechanical moment of inertia corresponding to free torsional vibrations is

$$J_{t,e} = \frac{m_1\{[35 + 2c_l(22 + 7c_l) + 87c_t + 6c_l(20 + 7c_l)c_t}{315(c_l + 2)^4}$$

$$+3(47 + 4c_l(18 + 7c_l))c_t^2 + (185 + 2c_l(370 + 7c_l(70$$

$$+c_l(40 + c_l(10 + c_l))))c_t^3]t_1^2 + 14(10 + 14c_l + 15c_l^2$$
(4.167)

$$+ (1 + c_l)(22 + c_l(4 + c_l)(11 + c_l(5 + c_l)))c_t)w^2\}$$

When $c_l \to 1$ (or $l_2 = l_1$) and $c_t \to 1$ (or $t_2 = t_1$) and $l_1 = l_2 = l/3$, this equation, too, reduces to Eq. (4.33), which stands for a constant-cross-section microbridge. The torsion-related resonant frequency can be calculated by Eqs. (4.166) and (4.167).

The configuration of Fig. 4.28 is a particular variant of the microbridge shown in Fig. 4.27, which is obtained by eliminating the middle portion. The relevant lumped-parameter resonant properties result from those describing the microbridge of Fig. 4.28, by taking c_l to be zero (which sets l_2 to be zero).

The bending stiffness of a long member is

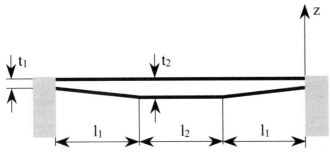

Figure 4.29 Another paddle microbridge with linearly variable thickness over the end segments.

$$k_{b,e} = \frac{E(c_t - 1)^3(c_t + 1)wt_1^3}{6l_1^3[2(1 - c_t) + (1 + c_t)\ln c_t]} \qquad (4.168)$$

For a short microbridge, the bending stiffness is

$$k_{b,e}^{sh} = \frac{2EG(c_t - 1)^3(c_t + 1)wt_1^3}{24G(c_t - 1)l_1^3 - (c_t + 1)l_1[12Gl_1^2}$$
$$+ \kappa E(c_t - 1)^2 t_1^2]\ln c_t \qquad (4.169)$$

The effective mass that corresponds to the free bending vibrations is

$$m_{b,e} = \frac{(63 + 193c_t)m_1}{315} \qquad (4.170)$$

The resonant bending frequency is, by way of Eqs. (4.168) and (4.170),

$$\omega_{b,e} = \frac{7.25(c_t - 1)t_1}{l_1^2}\sqrt{\frac{E(k_t^2 - 1)}{\rho(63 + 193c_t)[2(1 - c_t) + (1 + c_t)\ln c_t]}} \qquad (4.171)$$

The torsional stiffness is

$$k_{t,e} = \frac{4Gc_t^2 wt_1^3}{3(1 + c_t)l_1} \qquad (4.172)$$

The effective mechanical moment of inertia is

$$J_{t,e} = \frac{\{[35 + c_t(87 + c_t(141 + 185c_t))]t_1^2 + 28(5 + 11c_t)w^2\}m_1}{5040} \qquad (4.173)$$

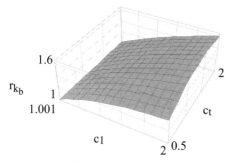

Figure 4.30 Comparison between the microbridges of Figs. 4.27 and 4.29 by means of the bending stiffness ratio.

The torsion-related resonant frequency is

$$\omega_{t,e} = \frac{81.98 c_t t_1}{l_1} \sqrt{\frac{G}{\rho(1 + c_t)\{[35 + c_t(87 + c_t(141 + 185 c_t))]t_1^2 + 28(5 + 11 c_t)w^2\}}} \quad (4.174)$$

Another paddle microbridge configuration is sketched in Fig. 4.29. The bending stiffness, which is associated to the midspan of this microbridge, is

$$k_{b,e} = \frac{16 E(c_l + c_t + c_t^2)(c_t - 1)^3 c_t^3 w t_1^3}{l_1^3[(c_t - 1)(c_l^4(c_t - 1)^2 + 24 c_l^2(c_t - 1)^2 c_t^2 - 192 c_t^4 + 4 c_l^3(c_t - 1)^2(c_t + 1)c_t + 48 c_l c_t^3(c_t - 3)) + 96 k_t^3(c_l + c_t + c_t^2)\ln c_t]} \quad (4.175)$$

The bending stiffness of a relatively short microcantilever is

$$k_{b,e}^{sh} = \frac{16 E G c_t^3(c_l + c_t + c_t^2)(c_t - 1)^3 w t_1^3}{l_1[G(c_t - 1)(c_l^4(c_t - 1)^2 + 24 c_l^2 c_t^2(c_t - 1)^2 - 192 c_t^4 + 4 c_l^3 c_t(c_t - 1)^2(c_t + 1) + 48 c_l c_t^3(c_t - 3))l_1^2 - 8 c_t^3(c_l + c_t + c_t^2)(12 G l_1^2 + \kappa E(c_t - 1)^2 t_1^2)\ln c_t]} \quad (4.176)$$

Example: A comparison is now made between the microbridges of Figs. 4.27 and 4.29 in terms of their bending stiffness, namely, by plotting the ratio of

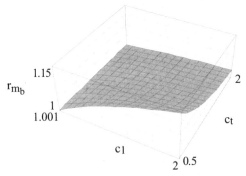

Figure 4.31   Comparison between the microbridges of Figs. 4.27 and 4.29 by means of the effective bending mass ratio.

the bending stiffness of the configuration shown in Fig. 4.27 [Eq. (4.164)] to the bending stiffness of the configuration sketched in Fig. 4.29 [Eq. (4.175)].

As Fig. 4.30 indicates, the stiffness corresponding to the microbridge of Fig. 4.27 can be 1.6 times larger than the stiffness describing the configuration of Fig. 4.29.

The effective mass corresponding to free bending vibrations is

$$m_{b,e} = \frac{128m_1\Big\{386 + 1300c_l + 1665c_l^2 + 960c_l^3 + 210c_l^4}{315(2 + c_l)^8}$$

$$+(1 + c_l)[\,126 + c_l(2 + c_l)(439 + c_l(813$$

$$+c_l(769 + c_l(351 + c_l(97 + c_l(15 + c_l))))))]c_t\Big\}}{315(2 + c_l)^8}$$

(4.177)

Another plot, shown in Fig. 4.31, which is similar to the previous ones, compares the two similar microbridge configurations in terms of their effective masses.

It can be seen that the differences between the effective masses of the two configurations are less marked, and that the maximum ratio is only 1.15. The resonant bending frequency, which is not given here due to its complex form, can simply be calculated by using the lumped-parameter stiffness [Eq. (4.175)] and mass [Eq. (4.177)].

The torsion stiffness, which is associated with the midspan of the microbridge shown in Fig. 4.29, is identical to the torsional stiffness of the microbridge sketched in Fig. 4.27, as expected, because the order in which the three component portions are interconnected is irrelevant in torsion. The same is not true with regard to the effective torsional inertias of the two compared configurations, which are different. The effective mechanical moment of inertia of the microbridge shown in Fig. 4.29 is

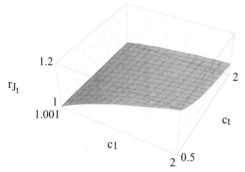

Figure 4.32   Comparison between the microbridges of Figs. 4.27 and 4.29 by means of the effective torsion mechanical moment of inertia ratio.

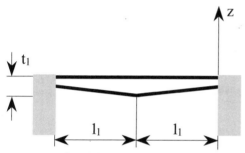

Figure 4.33   Simplified design of the paddle microbridge with linearly variable thickness over the end segments of Fig. 4.29.

$$J_{t,e} = \frac{m_1\{[5(37 + 4c_l(16 + 7c_l)) + 3(47 + 4c_l(18 + 7c_l))c_t + 3(29 + 2c_l(20 + 7c_l))c_t^2 + (35 + 2c_l(232 + 7c_l(61 + c_l(40 + c_l(10 + c_l)))))c_t^3]t_1^2 + 14[22 + 36c_l + 15c_l^2 + (1 + c_l)(10 + c_l(2 + c_l)(17 + c_l(7 + c_l)))c_t]w^2\}}{315(c_l + 2)^4} \quad (4.178)$$

As Fig. 4.32 indicates, the effective mechanical moment of inertia for the microbridge of Fig. 4.27 can be up to 1.2 times larger than the corresponding inertia of the configuration of Fig. 4.29.

The simplified version of the microbridge of Fig. 4.29 is the configuration shown in Fig. 4.33. Its lumped-parameter properties are derived by taking $c_l \to 0$ (which is the same as stating that $l_2 = 0$) in the equations describing the resonant characterization of the parent microbridge of Fig. 4.29. It can be shown that these lumped-parameter

Figure 4.34   Underneath three-dimensional view of a rectangular cross-section micro-bridge network.

properties are identical to those of the similar microbridge of Fig. 4.28 which are expressed in Eqs. (4.168) through (4.174).

4.5   Resonator Microbridge Arrays

Several microbridges having identical rectangular cross sections can be used in the same construction as a means of sensing a frequency range through resonance. Figure 4.34 shows such an assembly formed of several dissimilar-length microbridges.

The idea is to use a network of microbridges and microcantilevers which will resonantly cover a certain frequency range (the audible domain, from 20 to 20,000 Hz, for instance), as shown in the sketch of Fig. 4.34. Different sound pressures and frequencies will excite different resonant frequencies, and therefore one specific beam will resonate at a given input. It is thus possible to cover the entire audible range.

It can be shown that the bending resonant frequency of a fixed-fixed beam (bridge) is calculated as

$$l = 1.014 \frac{t}{f} \sqrt[4]{\frac{E}{\rho}}$$
(4.179)

Figure 4.35 plots this length for the audible range (up to 20,000 Hz), in the case of polysilicon beams with $E = 165$ GPa, $\rho = 2300$ kg/m^3, and for an average thickness of $t = 500$ nm, and this length profile should be identical to the geometric envelope shown in Fig. 4.34.

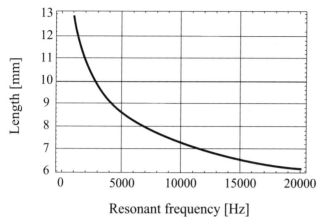

Figure 4.35   Length of constant-cross-section bridge in terms of resonant frequency.

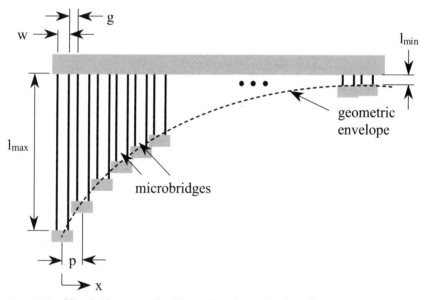

Figure 4.36   Microbridge network with nonlinearly varying length.

The relatively long bridges which are necessary at small resonant frequencies can be replaced by fixed-free beams (cantilevers) or by utilizing notches in a bridge; both solutions will enable us to obtain resonators of smaller lengths. The length profile of Fig. 4.35, which has been plotted based on Eq. (4.179), is also shown in Fig. 4.36.

Another microbridge network variant can be constituted of cylindrical wires, as sketched in Fig. 4.37.

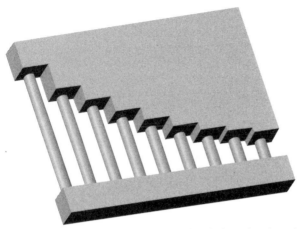

Figure 4.37   Underneath three-dimensional view of a wire microbridge network.

An equation similar to Eq. (4.179) gives the length of the wire in terms of a specific resonant frequency and the wire diameter, namely,

$$l = 0.943 \frac{t}{f} \sqrt[4]{\frac{E}{\rho}} \qquad (4.180)$$

An interesting aspect regards the precision of physically discretizing the continuous frequency spectrum that spans a specific range into a fixed number of stations. The finite width (or diameter) of the microbridge associated with a necessary gap between two consecutive members imposes the practical solution of utilizing a finite number of microbridges to cover a frequency range. The number of stations can simply be found by considering that both Eqs. (4.179) and (4.180) can be formulated to connect the length in terms of a distance x, shown in Fig. 4.36, instead of the frequency in the form:

$$l = \frac{c}{\sqrt{x}} \qquad (4.181)$$

If two limit lengths are selected, namely, $l_{\min}$ and $l_{\max}$, again shown in Fig. 4.36, then one can find the number of stations n as

$$n = \frac{x_{\max} - x_{\min}}{p} \qquad (4.182)$$

where $x_{\max}$ and $x_{\min}$ are found from Eq. (4.181) and p is the distance between the centers of two neighboring microbridges and is found based on the same Fig. 4.36 as

$$p = w + g \tag{4.183}$$

(For a circular cross-section microbridge, d should be employed instead of w.) In the end, the number of stations becomes

$$n = \frac{c^2}{w + g} \left(\frac{1}{l_{min}^2} - \frac{1}{l_{max}^2} \right) \tag{4.184}$$

for rectangular cross-section microbridges of widths w. For cylindrical wires, w is simply replaced by d in Eq. (4.184). The constant c is

$$c = 1.014 t \sqrt[4]{\frac{E}{\rho}} \tag{4.185}$$

for rectangular cross-section members and

$$c = 0.943 d \sqrt[4]{\frac{E}{\rho}} \tag{4.186}$$

for circular cross-section (wire) members.

References

1. B. Ilic, H. G. Craighead, S. Krylov, W. Senaratne, C. Ober, and P. Neuzil, Attogram detection using nanoelectromechanical devices, *Journal of Applied Physics*, **95**(7), 2004, pp. 3694–3703.

2. L. Sekaric, J. M. Parpia, H. G. Craighead, T. Feygelson, B. H. Houston, and J. E. Butler, Nanomechanical resonant structures in nanocrystalline diamond, *Applied Physics Letters*, **81**(23), 2002, pp. 4455–4456.

3. L. Sekaric, M. Zalatudinov, R. B. Bhiladvala, A. T. Zehnder, J. M. Parpia, and H. G. Craighead, Operation of nanomechanical resonant structures in air, *Applied Physics Letters*, **81**(14), 2002, pp. 2641–2643.

4. S. Evoy, A. Olkhovets, L. Sekaric, J. M. Parpia, H. G. Craighead, and D. W. Carr, Time-dependent internal friction in silicon nanoelectromechanical systems, *Applied Physics Letters*, **77**(15), 2000, pp. 2397–2399.

5. A. Husain, J. Home, H. W. C. Postma, X. M. H. Huang, T. Drake, M. Barbic, A. Scherer, and M. L. Roukes, Nanowire-based very-high-frequency electromechanical resonators, *Applied Physics Letters*, **83**(6), 2003, pp. 1240–1242.

6. R. R. A. Syms, Electrothermal frequency tuning of folded and coupled vibrating micromechanical resonators, *Journal of Microelectromechanical Systems*, **7**(2), 1998, pp. 164–171.

7. I. Dufour and L. Fadel, Resonant microcantilever type chemical sensors: Analytical modeling in view of optimization, *Sensors & Actuators B*, **91**, 2003, pp. 353–361.

8. F. Ayela and T. Fournier, An experimental study of anharmonic micromachined silicon resonators, *Measurement Science Technology*, **9**, 1998, pp. 1821–1830.

9. F. Plotz, S. Michaelis, R. Aigner, H.-J. Timme, J. Binder, and R. Noe, A low-voltage torsional actuator for application in RF-microswitches, *Sensors & Actuators A*, **92**, 2001, pp. 312–317.

10. Z. J. Yao, S. Chen, S. Eshelman, D. Denniston, and C. L. Goldsmith, Micromachined low-loss microwave switches, *Journal of Microelectromechanical Systems*, **8**, 1998, pp. 269–271.

11. D. Peroulis, S. Pacheco, and L. P. B. Katehi, MEMS devices for high isolation switching and tunable filtering, *IEEE MMT-S International Microwave Symposium Digest*, 2000, pp. 1217–1220.

12. J. Y. Park, G. H. Kim, K. W. Chung, and J. U. Bu, Monolithically integrated micromachined RF MEMS capacitive switches, *Sensors and Actuators A*, **89**, 2001, pp. 88–94.

13. G. M. Rebeiz, *RF MEMS: Theory, Design, and Technology*, Wiley Interscience, Hoboken, N. J., 2003.

14. M. Madou, *Fundamentals of Microfabrication*, 2d ed., CRC Press, Boca Raton, Fla., 2002.

15. S. Timoshenko, *Vibration Problems in Engineering*, D. Van Nostrand Company, New York, 1928.

16. S. S. Rao, *Mechanical Vibrations*, 2d ed., Addison-Wesley, Reading, Mass., 1990.

17. N. Lobontiu, *Compliant Mechanisms: Design of Flexure Hinges*, CRC Press, Boca Raton, Fla., 2002.

18. N. Lobontiu and E. Garcia, *Mechanics of Microelectromechanical Systems*, Kluwer Academic Press, New York, 2004.

Chapter

5

Resonant Micromechanical Systems

5.1 Introduction

Building upon the material presented thus far in the book, this chapter presents micromechanical resonators that are fabricated by integrating compliant parts with rigid and massive ones. Beam-type and spring-type resonators are first studied by using various modeling and design approaches which enable description of these mechanical resonators at different complexity levels. The main forms of actuation and sensing (transduction) that are currently implemented in micromechanical resonators are further analyzed, including electrostatic, electromagnetic, piezoelectric, piezomagnetic, and bimorph-based procedures. Real-life examples of microgyroscopes, tuning forks, and resonant sensors with the basic modeling apparatus and solved examples complete this chapter.

5.2 Beam-Type Microresonators

Several mechanical microresonators are studied in this section comprising flexible, beam-type members, which can deform by either bending or torsion. Three models are developed which are based on various levels of approximation with respect to stiffness and inertia properties and which yield the main resonant frequencies. The simplest resonant systems are single-degree-of-freedom, but multiple-degree-of-freedom (multiple-DOF) systems are also discussed. This study analyzes

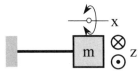

Figure 5.1   Paddle microcantilever with bending and torsional DOFs.

		Member # 1		Member # 2	
		Inertia	Stiffness	Inertia	Stiffness
Complexity	Model # I	●			●
	Model # II	●		●	●
	Model # III	●	●	●	●

Figure 5.2   Levels of model complexity in terms of lumped-parameter inertia and stiffness.

paddle microcantilevers, paddle microbridges, as well as other beam microresonators.

5.2.1 Resonant frequency models for microcantilevers

Paddle microcantilevers are discussed in Chap. 3 where their relevant lumped-parameter stiffness and inertia fractions are derived by considering contributions from both segments of the composite structure. A paddle microcantilever is sketched in Fig. 5.1, which highlights the main degrees of freedom (bending, represented by the z translation, and torsion, symbolized by the x axis rotation).

In some cases the end segment is massive and considerably stiffer than the root (anchor) one; these features warrant utilizing a simplified lumped-parameter model where stiffness only comes from the root segment whereas inertia is only produced by the free-end portion. An intermediate modeling case is also possible, resulting from the previous model, where inertia is contributed to by both segments. The third modeling possibility takes into consideration stiffness and inertia contributions from both segments and is thoroughly analyzed in Chap. 3. The three models are shown and characterized in Fig. 5.2.

While the stiffnesses yielded by models I and II are identical, the inertia provided by model II is larger than that of model I, and as a consequence, the resonant frequency of model I is always larger than that of model II.

Two paddle microcantilevers that were introduced in Chap. 3 are analyzed next by comparing the free bending and torsional responses provided by the three models. It should be mentioned that model III, the fully compliant, full-inertia model, was derived also in Chap. 3, and therefore only models I and II are formulated here.

Paddle microcantilever of constant thickness. The paddle microcantilever whose top view is shown in Fig. 3.8 and which is defined by two segments of constant thickness, and different widths and lengths, is studied now.

Model I.   According to model I of Fig. 5.2, the wider segment which is located at the free end in Fig. 3.8 is considered rigid whereas the root segment is the one providing compliance to the system. Likewise, the inertia contribution of only the free-end segment is taken into account.

The lumped-parameter bending stiffness comes from only the root segment and is

$$k_{b,e}^{\mathrm{I}} = \frac{Ew_2t^3}{4l_2^3} \qquad (5.1)$$

The mass of the free-end segment that corresponds to out-of-the-plane bending is

$$m_{b,e}^{\mathrm{I}} = \rho l_1 w_1 t \qquad (5.2)$$

and therefore the lumped-parameter bending resonant frequency yielded by model I becomes

$$\omega_{b,e}^{\mathrm{I}} = \sqrt{\frac{k_{b,e}^{\mathrm{I}}}{m_{b,e}^{\mathrm{I}}}} = \frac{t}{2l_2}\sqrt{\frac{Ew_2}{\rho l_1 l_2 w_1}} \qquad (5.3)$$

Similarly, the torsional stiffness of the root segment is identical to that of the whole microcantilever, and therefore

$$k_{t,e}^{\mathrm{I}} = \frac{Gw_2t^3}{3l_2} \qquad (5.4)$$

The torsional mechanical moment of inertia produced by the free-end segment is

$$J_{t,e}^{\mathrm{I}} = \frac{\rho l_1 w_1 t(w_1^2 + t^2)}{12} \qquad (5.5)$$

The torsional resonant frequency according to model I is therefore

$$\omega_{t,e}^{I} = \sqrt{\frac{k_{t,e}^{I}}{m_{t,e}^{I}}} = 2t\sqrt{\frac{Gw_2}{\rho l_1 l_2 w_1 (w_1^2 + t^2)}} \tag{5.6}$$

Model II. As suggested in Fig. 5.2, the next step in modeling complexity is model II, which considers that, in addition to the assumptions of the previous (simpler) model 1, inertia is produced by the root segment, and as a consequence, this is a full-inertia model.

For the paddle microcantilever of Fig. 3.8, the lumped-parameter bending-related inertia is

$$m_{b,e}^{II} = m_1 + \frac{33}{140}m_2 = \rho t\left(l_1 w_1 + \frac{33}{140}l_2 w_2\right) \tag{5.7}$$

The bending stiffness remains that given in Eq. (5.1). By combining Eqs. (5.1) and (5.7), the bending resonant frequency becomes

$$\omega_{b,e}^{II} = 0.5\frac{t}{l_2}\sqrt{\frac{Ew_2}{\rho l_2(l_1 w_1 + (33/140)l_2 w_2)}} \tag{5.8}$$

In torsion, the full inertia corresponding to free torsional vibrations is expressed as

$$J_{t,e}^{II} = J_{t1} + \frac{1}{3}J_{t2} = \frac{\rho t}{12}\left[l_1 w_1(w_1^2 + t^2) + \frac{l_2 w_2(w_2^2 + t^2)}{3}\right] \tag{5.9}$$

The resulting torsional resonant frequency can be determined by combining Eqs. (5.4) and (5.9) as

$$\omega_{t,e}^{II} = 2t\sqrt{\frac{Gw_2}{\rho l_2[l_1 w_1(w_1^2 + t^2) + l_2 w_2(w_2^2 + t^2)/3]}} \tag{5.10}$$

Example: Compare the bending and torsional resonant frequencies provided by the three models based on the example of a paddle microcantilever of constant thickness.

The following frequency ratios are formulated:

$$r\omega_b^{I-II} = \frac{\omega_{b,e}^{I}}{\omega_{b,e}^{II}} \qquad r\omega_b^{I-III} = \frac{\omega_{b,e}^{I}}{\omega_{b,e}^{III}} \tag{5.11}$$

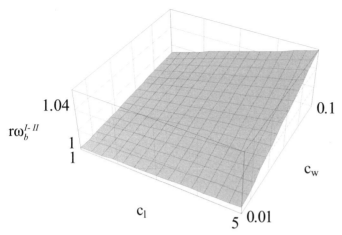

Figure 5.3 Bending resonant frequency ratio: model I predictions against model II predictions (paddle microcantilever of constant thickness).

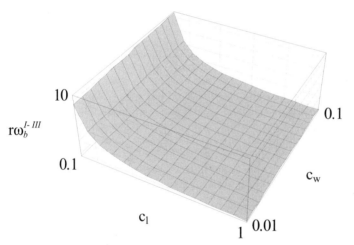

Figure 5.4 Bending resonant frequency ratio: model I predictions against model III predictions (paddle microcantilever of constant thickness).

where the superscripts I, II, and III indicate models I, II, and III, respectively. The bending resonant frequencies corresponding to models I and II are given in Eqs. (5.3) and (5.8), whereas the bending resonant frequency according to model III is formulated in Eq. (3.33). Figures 5.3 and 5.4 are three-dimensional plots of the ratios defined in Eqs. (5.11).

The following nondimensional parameters were used to draw Figs. 5.3 and 5.4:

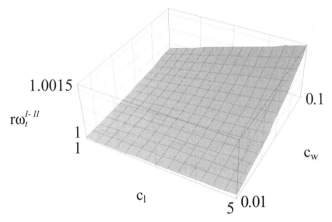

Figure 5.5 Torsion resonant frequency ratio: model I predictions against model II predictions (paddle microcantilever of constant thickness).

$$c_l = \frac{l_2}{l_1} \qquad c_w = \frac{w_2}{w_1} \tag{5.12}$$

The differences in predicting the bending resonant frequency by models I and II are negligible, as Fig. 5.3 indicates, for the parameter ranges selected. This indicates that by considering the inertia contribution from the root segment in addition to that of the end segment does not bring about substantial changes, and therefore models I and II are similar in predicting the bending resonant frequency. However, the differences between the predictions offered by models I and III are marked, quite severe at times, particularly for relatively short root segments, as Fig. 5.4 shows.

A similar comparison is made between the torsional resonant frequencies yielded by the three models for a paddle microcantilever. The following ratios are formulated:

$$r\omega_t^{I-II} = \frac{\omega_{t,e}^{I}}{\omega_{t,e}^{II}} \qquad r\omega_t^{I-III} = \frac{\omega_{t,e}^{I}}{\omega_{t,e}^{III}} \tag{5.13}$$

These enable us to draw the three-dimensional plots of Figs. 5.5 and 5.6 for the particular values of $w_1 = 80$ µm and $t = 0.8$ µm.

In torsion, too, the differences between the two model predictions of the resonant frequency are small, which indicates again that addition of the inertia effects of the root segment does not substantially alter the results provided by the simplest model I.

The situation changes when the comparison is performed between models I and III, as Fig. 5.6 suggests. The differences in the torsional resonant frequency values can be large, and for the selected parameter ranges, the

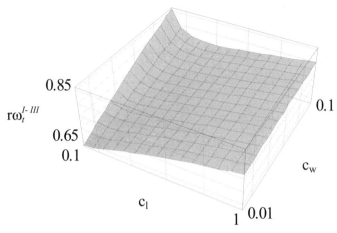

$r\omega_t^{I-III}$

0.85

0.65

0.1

0.1

c_w

c_1

1   0.01

Figure 5.6 Torsion resonant frequency ratio: model I predictions against model III predictions (paddle microcantilever of constant thickness).

predictions by the fully compliant, full-inertia model III are larger than those yielded by the simplified model I.

Paddle microcantilever of constant width. A study similar to that performed for the paddle microcantilever of constant thickness is now carried out for a paddle microcantilever of constant width, as the one introduced in Chap. 3 and sketched in Fig. 3.9. Resonant frequencies are again derived for bending and torsion according to the assumptions corresponding to models I and II defined in Fig. 5.2.

Model I. According to model I, inertia comes entirely from the free-end segment, whereas stiffness is furnished by only the thinner root segment. The lumped-parameter bending stiffness is therefore

$$k_{b,e}^{I} = \frac{Ewt_2^3}{4l_2^3} \tag{5.14}$$

The mass of the free-end segment is identical to the bending mass, according to the model I assumptions:

$$m_{b,e}^{I} = \rho l_1 w t_1 \tag{5.15}$$

The lumped-parameter bending resonant frequency can therefore be expressed as

$$\omega_{b,e}^{I} = \sqrt{\frac{k_{b,e}^{I}}{m_{b,e}^{I}}} = \frac{t_2}{2l_2}\sqrt{\frac{Et_2}{\rho l_1 l_2 t_1}} \tag{5.16}$$

The torsional stiffness of the root segment gives the stiffness of the entire microcantilever:

$$k_{t,e}^{I} = \frac{Gwt_2^3}{3l_2} \tag{5.17}$$

The torsional mechanical moment of inertia is

$$J_{t,e}^{I} = \frac{\rho l_1 w t_1 (w^2 + t_1^2)}{12} \tag{5.18}$$

The torsional resonant frequency combines the stiffness of Eq. (5.17) and the inertia fraction of Eq. (5.18) and is

$$\omega_{t,e}^{I} = \sqrt{\frac{k_{t,e}^{I}}{m_{t,e}^{I}}} = 2t_2\sqrt{\frac{Gt_2}{\rho l_1 l_2 t_1 (w^2 + t_1^2)}} \tag{5.19}$$

Model II. As previously specified, model II takes into consideration the inertia produced by the root segment, in addition to the assumptions of model I. The lumped-parameter bending-related inertia fraction is

$$m_{b,e}^{II} = m_1 + \frac{33}{140}m_2 = \rho w\left(l_1 t_1 + \frac{33}{140}l_2 t_2\right) \tag{5.20}$$

whereas the bending stiffness is that of model I, Eq. (5.14). The bending resonant frequency is

$$\omega_{b,e}^{II} = 0.5\frac{t_2}{l_2}\sqrt{\frac{Et_2}{\rho l_2(l_1 t_1 + (33/140)l_2 t_2)}} \tag{5.21}$$

The torsion-related inertia is

$$J_{t,e}^{II} = J_{t1} + \frac{1}{3}J_{t2} = \frac{\rho w}{12}\left[l_1 t_1(w^2 + t_1^2) + \frac{l_2 t_2(w^2 + t_2^2)}{3}\right] \tag{5.22}$$

The stiffness remains that formulated by model I, Eq. (5.17). The torsional resonant frequency is, by Eqs. (5.17) and (5.22),

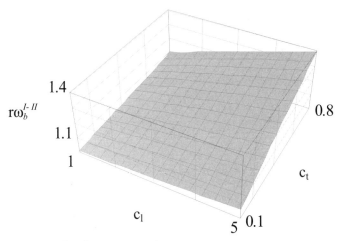

Figure 5.7 Bending resonant frequency ratio: model I predictions against model II predictions (paddle microcantilever of constant width).

$$\omega_{t,e}^{II} = 2t_2 \sqrt{\frac{Gt_2}{\rho l_2[l_1 t_1(w^2 + t_1^2) + l_2 t_2(w^2 + t_2^2)]/3]}} \qquad (5.23)$$

Example: Next we compare the bending and torsional resonant frequencies provided by the three models for a paddle microcantilever of constant width.

Equations (5.16) and (5.21) are utilized to form the bending resonant frequency ratio of the first of Eqs. (5.11), whereas Eqs. (5.16) and (3.41) are used to form the ratio of the second of Eqs. (5.11). Similarly, Eqs. (5.19) and (5.23) are employed to determine the torsional resonant frequency ratio of the first of Eqs. (5.13). Equations (5.19) and (3.38) are utilized in determining the ratio of the second of Eqs. (5.13). In all these ratios, the nondimensional parameter c_l defined in Eq. (5.12) is a variable together with

$$c_t = \frac{t_2}{t_1} \qquad (5.24)$$

As Fig. 5.7 shows, there are sensible differences between the bending resonant frequencies yielded by models I and II, and these differences increase linearly when the length and thickness of the root segment increase relative to their counterpart dimensions of the free-end segment. This divergent trend is even more marked between the bending resonant frequencies given by models I and III, as shown in Fig. 5.8, where the predictions by model I can be 8 times higher than those yielded by model III.

Figure 5.9 reveals a relationship between the torsional resonant frequencies obtained from models I and III, which is similar to that between the bending resonant frequencies of these models. When model I is compared to model III in terms of the torsional resonant frequency, it can be seen, as

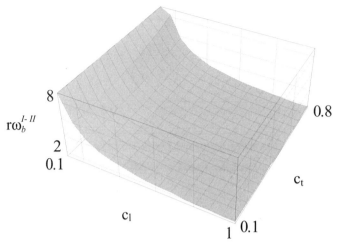

Figure 5.8   Bending resonant frequency ratio: model I predictions against model III predictions (paddle microcantilever of constant width).

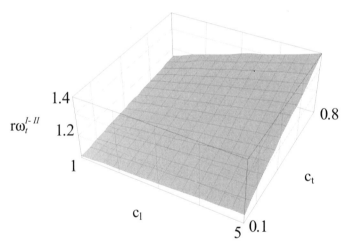

Figure 5.9   Torsion resonant frequency ratio: model I predictions against model II predictions (paddle microcantilever of constant width).

Fig. 5.10 indicates, that for relatively short and thin root segments, the resonant frequency predicted by model I is smaller than that of model III, whereas when c_l and c_t do increase, the relationship between the two model predictions reverses.

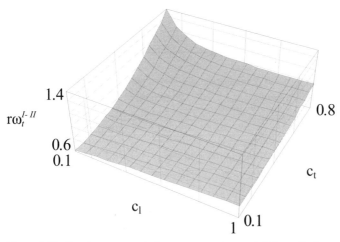

Figure 5.10 Torsion resonant frequency ratio: model I predictions against model III predictions (paddle microcantilever of constant width).

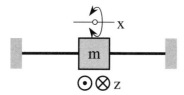

Figure 5.11 Schematic of paddle microbridge with its relevant degrees of freedom.

5.2.2 Resonant frequency models for microbridges

Microbridges are mostly utilized as micro- and nano-scale resonators enabling detection of deposition of extraneous substances through alteration of the resonant modes. The generic model of a paddle microbridge with its bending and torsional degrees of freedom highlighted is sketched in Fig. 5.11. Paddle microbridge designs, such as the one introduced in Chap. 4, consist of a middle segment which is placed at the structure's midpoint and two identical end segments, as shown in Fig. 4.19.

The three models of Fig. 5.2, which have been applied to paddle microcantilever configurations, can also be utilized in describing the resonant behavior of paddle microbridges. The simplest approach pertaining to model I of Fig. 5.2 takes into consideration the fact that the middle portion has larger dimensions and is usually assumed to be rigid, while the end segments are the ones ensuring the springiness of the whole member through their torsional and bending compliances. Often because of limitations imposed by the very small dimensions of

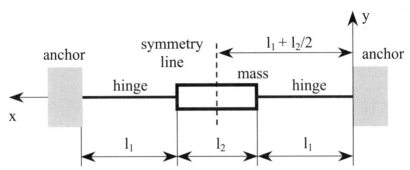

Figure 5.12 Paddle microbridge model with rigid body (mass) at the middle.

the paddle microbridge, as well as by a specific microfabrication process which yields thicknesses that are comparable in size for the compliant segments and the assumed-rigid segment (such as is the case with thin-film technologies), the elastic properties of the middle segment as well as the inertia of the end parts have to be accounted for in a model predicting the relevant resonant frequencies, as model III does.

The torsional and bending resonant frequencies are derived by assuming the middle segment is rigid, first by ignoring inertia contributions from the compliant parts (model I) and then by considering these inertia fractions (model II). The same resonant frequencies were determined in Chap. 4 by considering that the paddle microbridge is formed of fully compliant parts which all produce inertia fractions, and therefore contributions come in from all three segments in both the overall stiffness and effective inertia — these are the predictions of model III.

Model I.   The model of a paddle microbridge with a middle segment which is assumed rigid is shown in Fig. 5.12.

It can simply be shown that the bending stiffness produced by the two beam-springs that are in parallel is

$$k_{b,e}^{I} = \frac{24EI_{y1}}{l_1^3} = \frac{2Ew_1t^3}{l_1^3} \tag{5.25}$$

The mass is resulting from only the middle segment, which for the configuration of Fig. 5.12 is

$$m_{b,e}^{I} = \rho l_2 w_2 t \tag{5.26}$$

The bending resonant frequency is

$$\omega_{b,e}^{\mathrm{I}} = \sqrt{\frac{k_{b,e}^{\mathrm{I}}}{m_{b,e}^{\mathrm{I}}}} = \sqrt{2}\,\frac{t}{l_1}\sqrt{\frac{Ew_1}{\rho w_2 l_1 l_2}} \tag{5.27}$$

As a reminder, the assumptions are made in Eqs. (5.25), (5.26), and (5.27) that the three segments have identical thicknesses t and that their widths are different: w_1 for the end segments and w_2 for the middle one ($w_2 > w_1$).

The torsional stiffness associated with the midpoint of the paddle microbridge is again a combination of the two beam-springs, namely,

$$k_{t,e}^{\mathrm{I}} = \frac{2GI_{t1}}{l_1} = \frac{2Gw_1 t^3}{3l_1} \tag{5.28}$$

and the torsional moment of inertia of the paddle microbridge is identical to that of the middle segment according to the assumptions of model I, namely,

$$J_{t,e}^{\mathrm{I}} = \frac{\rho l_2 w_2 t(w_2^2 + t^2)}{12} \tag{5.29}$$

The torsional resonant frequency can now be calculated as

$$\omega_{t,e}^{\mathrm{I}} = \sqrt{\frac{k_{t,e}^{\mathrm{I}}}{J_{t,e}^{\mathrm{I}}}} = 2\sqrt{2}\,t\sqrt{\frac{Gw_1}{\rho w_2 l_1 l_2(w_2^2 + t^2)}} \tag{5.30}$$

Model II. Model I ignored the inertia contributions of the end compliant segments, and this is acceptable as long as the mass of the middle segment is (considerably) larger than the masses of the end segments. However, in cases where the two different segments comprising the paddle bridge are comparable, the compliant end segments also contribute to the total mass, and this additional mass can be calculated by Rayleigh's principle, which states that the deformation (displacement) distribution of a beam in this case is identical to the velocity distribution of the same member. By equating the kinetic energy of the real, distributed-parameter beam to that of the equivalent lumped-parameter system, the additional mass is calculated as

$$m_{1,e} = \rho t w_1 \int_0^{l_1} f_b^2(x)\,dx \tag{5.31}$$

For a fixed-guided beam (as the root segment of the paddle microbridge is), the distribution function is

$$f_b(x) = \frac{(l_1 - x)^2(l_1 + 2x)}{l_1^3} \tag{5.32}$$

By substituting the distribution function of Eq. (5.32) into Eq. (5.31), the effective mass corresponding to a root segment becomes

$$m_{1,e} = \frac{13}{35} m_1 = \frac{13}{35} \rho l_1 w_1 t \tag{5.33}$$

and therefore the total mass undergoing free bending vibrations is

$$m_{b,e}^{II} = 2\frac{13}{35} m_1 + m_2 = \rho t (l_2 w_2 + \frac{26}{35} l_1 w_1) \tag{5.34}$$

The bending resonant frequency can therefore be expressed as

$$\omega_{b,e}^{II} = \sqrt{\frac{k_{b,e}^{II}}{m_{b,e}^{II}}} = \sqrt{70}\frac{t}{l_1}\sqrt{\frac{Ew_1}{\rho l_1(26 l_1 w_1 + 35 l_2 w_2)}} \tag{5.35}$$

The effective (equivalent) torsional mechanical moment of inertia of one end microbar is

$$J_{t1,e} = \frac{1}{3} m_1 \frac{w_1^2 + t^2}{12} = \frac{1}{3} J_{t1} \tag{5.36}$$

and therefore the total torsion-related inertia fraction is

$$J_{t,e}^{II} = 2J_{t1,e} + J_{t2,e} = \frac{\rho t}{12}\left[l_2 w_2(w_2^2 + t^2) + \frac{2 l_1 w_1}{3}(w_1^2 + t^2) \right] \tag{5.37}$$

The resulting torsional resonant frequency can be expressed as

$$\omega_{t,e}^{II} = \sqrt{\frac{k_{t,e}^{II}}{J_{t,e}^{II}}} = 2\sqrt{6}t\sqrt{\frac{Gw_1}{\rho l_1[2 l_1 w_1(w_1^2 + t^2) + 3 l_2 w_2(w_2^2 + t^2)]}} \tag{5.38}$$

Figure 5.13 illustrates the relative errors in the bending resonant frequency that are generated between the predictions of model I [Eqs. (5.27)] and model II [Eqs. (5.35)] as a function of the parameters defined in Eqs. (5.12). When the dimension parameters of the two different segments c_l and c_w are close to 1, the errors are larger; but they

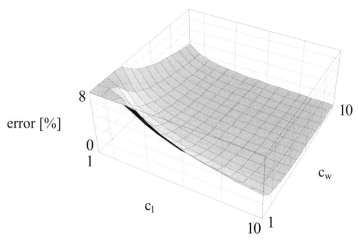

Figure 5.13 Percentage error in the bending resonant frequency: model I predictions against model II predictions (paddle microbridge).

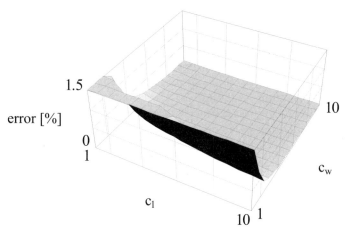

Figure 5.14 Percentage error in the torsional resonant frequency: model I predictions against model II predictions (paddle microbridge).

diminish as the dimensions of the middle segment increase compared to those of the end ones.

The relative errors between the torsional resonant frequencies—Eqs. (5.30) versus Eq. (5.38)—are plotted in Fig. 5.14 in terms of the nondimensional parameters defined in Eq. (5.12), and for a baseline geometry of $t = 1$ µm and $w_1 = 5$ µm.

For the selected parameter ranges and baseline parameters, the errors in the torsional resonant frequency when ignoring the inertia contributions from the two compliant end segments are smaller than

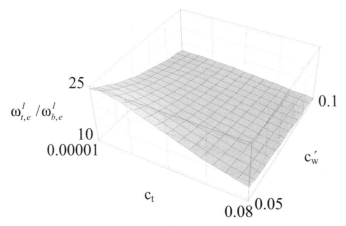

Figure 5.15 Torsion-to-bending resonant frequency ratio according to model I predictions (paddle microbridge).

those generated in the case of the bending resonant frequency. These errors, too, tend to decrease with the relative increase in the middle-segment dimensions.

A comparison is made now between the bending and resonant frequencies, by using the torsion-to-bending resonant frequency ratio—Eqs. (5.30) and (5.27), which do not take into account inertia fractions from the end compliant segments. Figure 5.15 is the three-dimensional plot of the torsion-to-bending resonant frequency ratio, and it is plotted as a function of the following parameters:

$$c_t = \frac{t}{l_1} \qquad c'_w = \frac{w_2}{l_1} \qquad (5.39)$$

A similar comparison is illustrated in Fig. 5.16, which uses Eqs. (5.38) and (5.35) in comparing the two relevant resonant frequencies, and therefore takes into consideration inertia contributions from the end compliant segments. It can be seen from Figs. 5.15 and 5.16 that the model predicts higher torsional resonant frequency values for the parameter ranges selected.

A comparison is also made between the torsional resonant frequency [Eq. (4.130)] and the bending resonant frequency [Eqs. (4.131) and (4.132) according to the fully compliant, full-inertia model III], and Fig. 5.17 is the three-dimensional plot illustrating it.

Predictions by the fully compliant segments model are similar to those made by the model with the rigid middle segment, in the sense that the torsional resonant frequency is higher than the bending one.

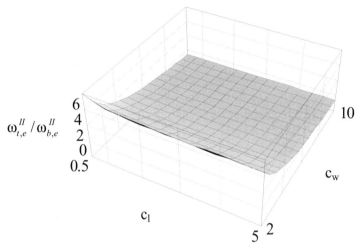

$\omega_{t,e}^{II} / \omega_{b,e}^{II}$

Figure 5.16 Torsion-to-bending resonant frequency ratio according to model II predictions (paddle microbridge).

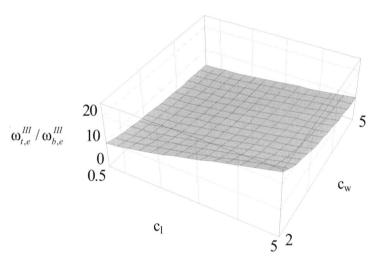

$\omega_{t,e}^{III} / \omega_{b,e}^{III}$

Figure 5.17 Torsion-to-bending resonant frequency according to model III predictions (paddle microbridge).

Model I and Model III can also be compared by studying their resonant responses in bending and torsion in a manner similar to that used to compare the predictions by the first two models for paddle microcantilevers. Figure 5.18 is the three-dimensional plot of the bending resonant frequency ratio, whereas Fig. 5.19 is the plot of the torsional resonant frequency.

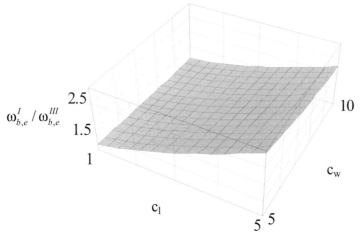

Figure 5.18 Bending resonant frequency ratio: model I predictions against model III predictions (paddle microbridge).

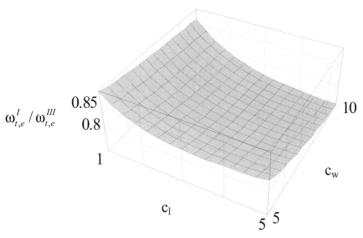

Figure 5.19 Torsion resonant frequency ratio: model I predictions against model III predictions (paddle microbridge).

Compared to the paddle microbridge where the subscript 1 indicated the wider segment, the subscript 1, as also indicated in Fig. 5.18, denotes the narrower, root segment, and therefore the values of parameters c_w and c_l have been chosen accordingly. As Fig. 5.18 shows, the bending resonant frequency prediction by model I (the simplest model) is larger than the prediction yielded by model III (the fully compliant, full-inertia model). Figure 5.19 has been drawn for the following particular parameter values: $l_1 = 100$ µm, $w_1 = 20$ µm, and $t_1 = 0.5$ µm. For

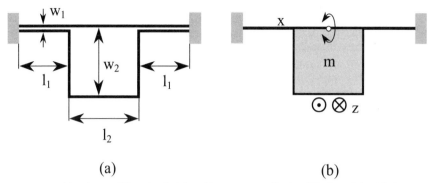

(a)   (b)

Figure 5.20   Torsion paddle microbridge: (a) geometry; (b) simplified model with degrees of freedom.

these values and the ranges allotted to the nondimensional parameters c_l and c_w, Fig. 5.19 indicates that the torsional resonant frequency predicted by model I is smaller than that produced by model III.

5.2.3   Other examples of beam-type microresonators

Several other examples of microresonators are discussed next that are designed by using relatively large segments which can be considered rigid, connected to smaller components which are flexible. As a consequence, all following resonators will be qualified according to the assumptions of model I, where the inertia fraction was only produced by the massive segment whereas the stiffness was given by the relatively smaller compliant segments.

Example: Analyze the torsional resonant frequency of the torsional bridge sketched in Fig. 5.20a and compare it to the corresponding resonant frequency of the regular bridge design illustrated in Fig. 4.19.

The design of Fig. 5.20a is analyzed by Xiao et al.[1] as an electrostatic torsional actuator and by Selvakumar and Najafi[2] as a microaccelerometer, but it can also be utilized as a torsional resonator for mass addition detection purposes—the subject of mass addition detection is studied more thoroughly in Chap. 6. The assumption made here is that the central segment is rigid (and possesses a mass m, as indicated in Fig. 5.20b), whereas the identical root segments are compliant and are considerably lighter than the middle segment such that their inertia contribution can be neglected.

Compared to the paddle microbridge of Fig. 4.19, the design sketched in Fig. 5.20a is quite similar except for the middle segment which is placed asymmetrically with respect to the longitudinal axis of the hinges. Figure 5.20b indicates the main possible motions (degrees of freedom) which are the rotation around the x axis (implying torsion of the root segments) and out-of-the-plane translation about the z axis (which is achieved through bending of the root flexible segments about the y axis).

The torsional stiffness is provided by the two root segments and is explicitly given in Eq. (5.28). The mechanical torsional moment of inertia is produced by only the middle segment. When we consider that this segment is a relatively thin plate, its moment of inertia about its central (symmetry) axis is, as indicated by Beer and Johnston[3] for instance,

$$J_{t,C} = \frac{m w_2^2}{12} \tag{5.40}$$

Because of the fact that the bridge system of Fig. 5.20b will rotate about the x axis, the moment of inertia about that axis will be, according to the parallel axes theorem,

$$J_{t,e} = \frac{m w_2^2}{12} + \frac{m w_2^2}{4} = \frac{m w_2^2}{3} \tag{5.41}$$

The torsional resonant frequency of this microbridge becomes

$$\omega_{t,e} = \sqrt{2} \frac{t}{w_2} \sqrt{\frac{G w_1}{\rho l_1 l_2 w_2}} \tag{5.42}$$

By using the nondimensional parameters

$$c_w = \frac{w_2}{w_1} \qquad c_t' = \frac{t}{w_1} \tag{5.43}$$

the following torsional frequency ratio can be formulated:

$$\frac{\omega_{t,e}^*}{\omega_{t,e}} = \frac{2}{\sqrt{1 + (c_t'/c_w)^2}} \tag{5.44}$$

where $\omega_{t,e}^*$ is the torsional resonant frequency of the symmetric paddle microbridge shown in Fig. 4.19. Figure 5.21 is the three-dimensional plot of the resonant frequency ratio expressed in Eq. (5.44). It can be seen that for the selected parameter ranges this frequency ratio remains almost constant and equal to 2, which indicates that the torsional resonant frequency of the symmetric paddle bridge (Fig. 4.19) is twice the frequency of the asymmetric bridge shown in Fig. 5.20a.

Example: Study the free response of the four-arm microbridge sketched in Fig. 5.22, and compare it to the resonant behavior of a regular symmetric microbridge such as the one shown in Fig. 4.19. Assume the central segment is rigid and the legs are massless and compliant. Also assume the geometry

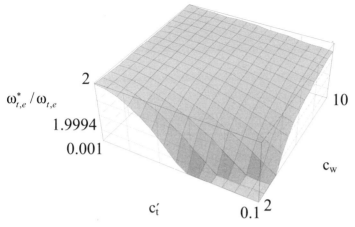

$$\omega_{t,e}^{*} / \omega_{t,e}$$

Figure 5.21   Torsional frequency ratio: comparison between symmetric and asymmetric microbridge designs.

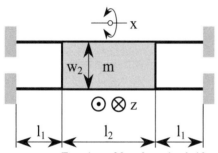

Figure 5.22   Top view of four-leg microbridge.

of the four-leg design is identical to that of the paddle microbridge of Fig. 4.19. This microdevice was studied by Degani et al.,[4] among others, as a vibrating rate gyroscope.

Figure 5.23 is a simplified mechanical model which considers the inertia of the middle plate and the stiffness contributions produced by the four compliant legs in connection to the 2 degrees of freedom of the middle plate, which are the translation u_z and the rotation θ_x. Translation about the z direction is realized by bending of the four compliant legs, and the linear stiffness of such a fixed-free beam is

$$k_l = \frac{E w_1 t_1^3}{4 l_1^3} \tag{5.45}$$

Rotation of the middle plate about its central (symmetry) axis, shown in Fig. 5.23, is mainly realized through torsion of the four root compliant legs, and therefore the torsional stiffness of such a bar is

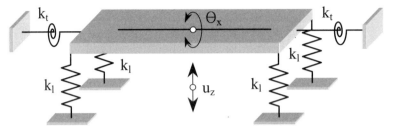

Figure 5.23 Four-leg microbridge: lumped-parameter model and degrees of freedom.

$$k_t = \frac{Gw_1 t_1^3}{3l_1} \tag{5.46}$$

The mass of the whole microstructure comes only from the middle plate, which is

$$m = \rho l_2 w_2 t_2 \tag{5.47}$$

and the mechanical moment of inertia of the same plate with respect to the x axis shown in Fig. 5.23 is

$$J_C = \frac{m w_2^2}{12} \tag{5.48}$$

The assumption is made in this model that the thickness of the identical compliant legs t_1 is different from the thickness of the middle plate t_2.

To derive the free response in matrix form for this 2-DOF model, Lagrange's equations are employed. The kinetic energy of the microsystem is produced by z translation and x rotation of the middle plate, which is assumed rigid and the only component contributing to overall inertia. Its expression is

$$T = \frac{1}{2} m \dot{u}_z^2 + \frac{1}{2} J_C \dot{\theta}_x^2 \tag{5.49}$$

Similarly, the potential energy of the microsystem is produced through elastic bending and torsion deformations of the four identical root components, namely,

$$U = 2 \left[\frac{1}{2} k_l \left(u_z + \frac{w_2}{2\theta_x} \right)^2 + \frac{1}{2} k_l \left(u_z - \frac{w_2}{2\theta_x} \right)^2 \right] + 4 \frac{1}{2} k_t \theta_x^2 \tag{5.50}$$

Equation (5.50) took into account the fact that due to superimposed translation and rotation of the middle platform, two sets of springs deform differently in bending. For two of them the total deflection is $u_z + \theta_x w_2/2$, whereas for the other two the total deflection is $u_z - \theta_x w_2/2$.

By applying Lagrange's equations the following differential equations system is obtained:

$$m\ddot{u}_z + 4k_l u_z = 0 \qquad J_C\ddot{\theta}_x + (4k_t + k_l w_2^2)\theta_x = 0 \tag{5.51}$$

which indicates that the two motions are dynamically decoupled, according to previous definitions introduced in Chap. 1. The mass and stiffness matrices corresponding to Eqs. (5.51) are

$$[M] = \begin{bmatrix} m & 0 \\ 0 & J_C \end{bmatrix} \qquad [K] = \begin{bmatrix} 4k_l & 0 \\ 0 & 4k_r + k_l w_2^2 \end{bmatrix} \tag{5.52}$$

The dynamic matrix can now be formed as

$$[A] = [M]^{-1}[K] = \begin{bmatrix} \dfrac{4k_l}{m} & 0 \\ 0 & \dfrac{4k_r + k_l w_2^2}{J_C} \end{bmatrix} \tag{5.53}$$

and its eigenvalues yield the following resonant frequencies:

$$\omega_1 = \frac{2t_1}{l_1 w_2}\sqrt{\frac{t_1 w_1(16Gl_1^2 + 3Ew_2^2)}{\rho l_1 l_2 t_2 w_2}} \qquad \omega_2 = \frac{t_1}{l_2}\sqrt{\frac{Et_1 w_1}{\rho l_1 l_2 t_2 w_2}} \tag{5.54}$$

In the particular case where the thickness of the middle plate is equal to that of the four identical root legs, Eqs. (5.54) simplify to

$$\omega_1 = \frac{2t}{l_1 w_2}\sqrt{\frac{w_1(16Gl_1^2 + 3Ew_2^2)}{\rho l_1 l_2 w_2}} \qquad \omega_2 = \frac{t}{l_2}\sqrt{\frac{Ew_1}{\rho l_1 l_2 w_2}} \tag{5.55}$$

Equations (5.54) and (5.55) indicate that the resonant frequency ω_1 combines both torsion and bending effects (through the simultaneous presence of the elastic modulii E and G), whereas the other resonant frequency ω_2 contains only the bending contribution. The combined effect of torsion and bending in the resonant frequency ω_1 is normal as rotation of the middle plate around its central x axis results through the mixed torsion-bending deformation of the four root legs.

Example: Compare the resonant frequencies of the four-leg microbridge sketched in Fig. 5.22.

The following resonant frequency ratio can be formulated by using Eqs. (5.54) and (5.55):

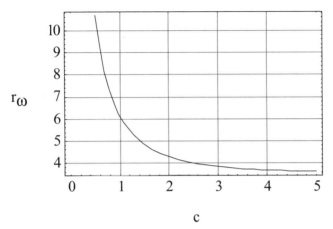

Figure 5.24 Resonant frequency ratio [Eq. (5.56)] in terms of the nondimensional parameter c [Eq. (5.57)] when the root legs are assumed fixed-free.

$$r_\omega = \frac{\omega_1}{\omega_2} = 2\sqrt{3 + \frac{G}{E}\left(\frac{4}{c}\right)^2} \tag{5.56}$$

where

$$c = \frac{w_2}{l_1} \tag{5.57}$$

By using the regular relationship between the elastic modulii

$$G = \frac{E}{2(1+\mu)} \tag{5.58}$$

and for a value of $\mu = 0.25$ (corresponding to polysilicon material), the resonant frequency ratio of Eq. (5.56) is plotted in Fig. 5.24 against the nondimensional parameter c of Eq. (5.57).

As Fig. 5.24 illustrates, the resonant frequency ω_1 is at least 3 times larger than the resonant frequency ω_2, and as the ratio c of Eq. (5.57) increases, the resonant frequency ratio decreases. When $w_2 = l_1$, for instance, ω_1 is 6 times larger than ω_2, whereas when $w_2 = 2.5 l_1$ ($c = 2.5$), ω_1 is only 4 times larger than ω_2.

All this example's derivation was based on the simplifying assumption that the four root legs are fixed at their anchor end and free at their connection to the plate. A closer look at the boundary conditions and the 2 degrees of freedom of the plate indicate that the translatory-rotary motion combination demands that the connection point between a root leg and the plate be guided. As such, the linear stiffness of a fixed-guided root leg is 4 times larger than the stiffness of a fixed-free leg, namely,

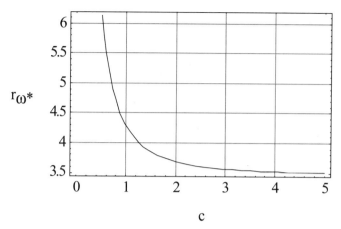

Figure 5.25 Resonant frequency ratio [Eq. (5.61)] in terms of the nondimensional parameter c [Eq. (5.57)] when the root legs are assumed fixed-guided.

$$k_l^* = \frac{E w_1 t_1^3}{l_1^3} \qquad (5.59)$$

By carrying out all the previous calculations, it follows that the new resonant frequencies are

$$\omega_1^* = \frac{4 t_1}{l_1 w_2} \sqrt{\frac{t_1 w_1 (4 G l_1^2 + 3 E w_2^2)}{\rho l_1 l_2 t_2 w_2}} \qquad \omega_2^* = 2 \omega_2 \qquad (5.60)$$

under the assumption that the root legs and middle plate have different thicknesses, and in the case where the same thickness is used all over, Eqs. (5.60) can simply be used with $t_1 = t_2 = t$ to get the corresponding resonant frequencies. The following resonant frequency ratio can be formulated:

$$r\omega^* = \frac{\omega_1^*}{\omega_2^*} = 2 \sqrt{3 + \frac{G}{E} \left(\frac{2}{c}\right)^2} \qquad (5.61)$$

and Fig. 5.25 plots this ratio in terms of the nondimensional parameter c of Eq. (5.57).

Figure 5.25 reveals similar trends to those discussed with respect to Fig. 5.24 where the four legs were assumed fixed-free. The descent in the resonant frequency ratio curve is steeper in Fig. 5.25 than the similar descent in Fig. 5.24, but the curves in the two figures tend toward the same limit of $2\sqrt{3} \approx 3.5$ when the parameter c increases.

Example: Analyze the resonant motions of the microstructure, illustrated in Fig. 5.26, which is composed of four identical legs that are placed radially

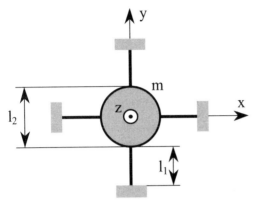

Figure 5.26   Top view of resonator with four identical legs and central plate.

symmetrically with respect to the central plate. Assume the plate is rigid and the legs are compliant and that inertia comes from only the plate. A similar design was analyzed by Ayela and Fournier,[5] for instance.

The shape of the plate in Fig. 5.26 is circular, but it could also be of a different shape, provided it possesses two symmetry axes (the x and y axes). In such cases, two different dimensions should define its planar envelope instead of l_2, which characterizes the plate in Fig. 5.26. The derivation in the case of a plate with two planar dimensions is not pursued here, but it can simply be derived from the current presentation.

The central plate is capable of the following independent motions: translation about the z axis, rotation about the z axis, and rotations about the x and y axes. As a consequence, the resonator of Fig. 5.26 is defined by 4 degrees of freedom, namely, u_z, θ_z, θ_x, and θ_y. The Lagrange approach and associated equations are again utilized to formulate the equations of free motion and further evaluate the system's resonant frequencies.

The kinetic energy is

$$T = \frac{1}{2}m\dot{u}_z^2 + \frac{1}{2}J_z\dot{\theta}_z^2 + \frac{1}{2}J_C\dot{\theta}_x^2 + \frac{1}{2}J_C\dot{\theta}_y^2 \tag{5.62}$$

where J_z is the mechanical moment of inertia of the plate with respect to the z axis and is given by

$$J_z = \frac{ml_2^2}{8} \tag{5.63}$$

and J_C is the mechanical moment of inertia of the plate with respect to either the x or the y axis passing through the symmetry center, namely,

$$J_C = \frac{ml_2^2}{16} \tag{5.64}$$

The expressions of Eqs. (5.63) and (5.64) can be found in Beer and Johnston,[3] for instance.

The potential energy of the spring system corresponding to the resonator of Fig. 5.26 can be expressed as

$$U = U_{b,x} + U_{b,y} + U_{b,z} + U_t \qquad (5.65)$$

where $U_{b,x}$, $U_{b,y}$, and $U_{b,z}$ are elastic energy terms produced through bending of various legs about the x, y, and z axis, respectively, whereas U_t is the elastic energy produced through torsion of the four legs. The four different potential energy terms of Eq. (5.65) are expressed next. The $U_{b,x}$ potential energy is

$$U_{b,x} = \frac{1}{2}k_{l,o}\left(u_z + \frac{l_2}{2}\theta_x\right)^2 + \frac{1}{2}k_{l,o}\left(u_z - \frac{l_2}{2}\theta_x\right)^2 \qquad (5.66)$$

where $k_{l,o}$ is the linear stiffness of a leg corresponding to out-of-the-plane bending and is expressed for fixed-free boundary conditions as

$$k_{l,o} = \frac{Ew_1t_1^3}{4l_1^3} \qquad (5.67)$$

The $U_{b,y}$ potential energy is similarly expressed as

$$U_{b,y} = \frac{1}{2}k_{l,o}\left(u_z + \frac{l_2}{2}\theta_y\right)^2 + \frac{1}{2}k_{l,o}\left(u_z - \frac{l_2}{2}\theta_y\right)^2 \qquad (5.68)$$

The $U_{b,z}$ potential energy term is calculated as

$$U_{b,z} = 4\left[\frac{1}{2}k_{l,i}\left(\frac{l_2}{2}\theta_z\right)^2\right] \qquad (5.69)$$

where $k_{l,i}$ is the linear stiffness of a root leg corresponding to the in-plane bending and is given by

$$k_{l,i} = \frac{Ew_1^3t_1}{4l_1^3} \qquad (5.70)$$

Eventually, the torsional potential energy term of Eq. (5.65) is

$$U_t = 2\left(\frac{1}{2}k_t\theta_x^2 + \frac{1}{2}k_t\theta_y^2\right) \qquad (5.71)$$

By applying Lagrange's equations, the differential equations which govern the motion of the plate of Fig. 5.26 are

$$m\ddot{u}_z + 4k_{l,o}u_z = 0$$

$$J_C\ddot{\theta}_x + \frac{1}{2}(4k_t + k_{l,o}l_2^2)\theta_x = 0$$

$$J_C\ddot{\theta}_y + \frac{1}{2}(4k_t + k_{l,o}l_2^2)\theta_y = 0 \qquad (5.72)$$

$$J_z\ddot{\theta}_z + k_{l,i}l_2^2\theta_z = 0$$

The mass matrix corresponding to the equations system (5.72) is

$$[M] = \begin{bmatrix} m & 0 & 0 & 0 \\ 0 & J_C & 0 & 0 \\ 0 & 0 & J_C & 0 \\ 0 & 0 & 0 & J_z \end{bmatrix} \qquad (5.73)$$

The stiffness matrix of the same system (5.72) is

$$[K] = \begin{bmatrix} 4k_{l,o} & 0 & 0 & 0 \\ 0 & \frac{1}{2}(4k_t + k_{l,o}l_2^2) & 0 & 0 \\ 0 & 0 & \frac{1}{2}(4k_t + k_{l,o}l_2^2) & 0 \\ 0 & 0 & 0 & k_{l,i}l_2^2 \end{bmatrix} \qquad (5.74)$$

The dynamic matrix [A] can now be formulated, and the three corresponding eigenvalues produce the following resonant frequencies:

$$\omega_1 = l_2\sqrt{\frac{k_{l,i}}{J_z}} \qquad \omega_2 = \omega_3 = \sqrt{\frac{4k_t + k_{l,o}l_2^2}{2J_c}} \qquad \omega_4 = 2\sqrt{\frac{k_{l,o}}{m}} \qquad (5.75)$$

Equations (5.75) show that there are three distinct modes, as indicated by the three distinct resonant frequencies. One mode, described by the first of Eqs. (5.75), consists in the z rotation of the plate and is produced by in-plane bending of the root compliant segments. Two identical modes, defined by the second of Eqs. (5.75), are oscillations of the plate around either the x or the y axis and are produced by combined torsion and out-of-the-plane bending of the root segments. Eventually, one last mode is defined by the third of Eqs. (5.75) and consists of a translatory (piston-type) motion of the plate about the z axis which is generated through out-of-the-plane bending of the compliant segments.

The following resonant frequency ratios can be formulated:

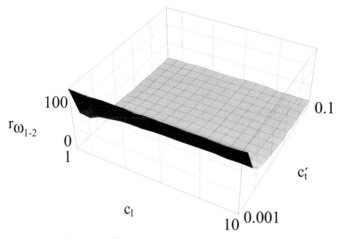

Figure 5.27   Resonant frequency ratio, the first of Eqs. (5.76).

$$r\omega_{1-2} = \frac{\omega_1}{\omega_2} = \sqrt{3}\frac{l_2 w_1}{t_1}\sqrt{\frac{E}{16Gl_1^2 + 3El_2^2}}$$

$$r\omega_{2-4} = \frac{\omega_2}{\omega_4} = \sqrt{1 + \frac{16}{3}\frac{G}{E}\left(\frac{l_1}{l_2}\right)^2}$$
(5.76)

By using the nondimensional variables

$$c_l = \frac{l_2}{l_1} \qquad c_t' = \frac{t_1}{w_1}$$
(5.77)

the three-dimensional plot of Fig. 5.27 and the two-dimensional plot of Fig. 5.28 are obtained.

As Fig. 5.27 indicates, the resonant frequency ω_1 can be 100 times larger than the next resonant frequency ω_2, especially when the nondimensional parameter c_t' (to which the frequency ratio is particularly sensitive) is small. The last resonant frequency is the smallest of the three, as indicated in Fig. 5.28, which also shows that the resonant frequencies ω_2 and ω_4 tend to be equal for large values of the plate dimension l_2.

Example: Determine the resonant frequency of the system shown in Fig. 5.29 by considering only the two rotation DOFs (the ones produced through torsion). Harrington, Mohanty, and Roukes[6] utilized this design to study energy dissipation issues in micromechanical resonators.

The microdevice of Fig. 5.29 can be modeled as a 2-DOF system, the generalized coordinates being the two rotation angles θ_{1x} and θ_{2x}. The kinetic energy of the system is

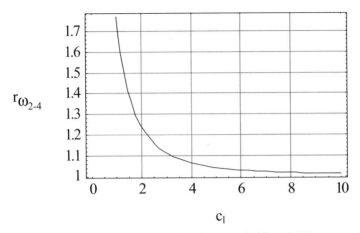

Figure 5.28 Resonant frequency ratio, the second of Eqs. (5.76).

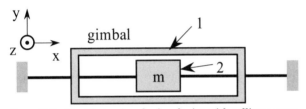

Figure 5.29 Double torsional microdevice with collinear axes.

$$T = \frac{1}{2} J_1 \dot{\theta}_{1x}^2 + \frac{1}{2} J_2 \dot{\theta}_{2x}^2 \tag{5.78}$$

and the potential energy which is stored in the two pairs of torsional springs is

$$U = k_{t1} \theta_{1x}^2 + k_{t2}(\theta_{1x} - \theta_{2x})^2 \tag{5.79}$$

The following dynamic equations result from applying Lagrange's equations based on the energies defined in Eqs. (5.78) and (5.79):

$$J_1 \ddot{\theta}_{1x} + 2(k_{t1} + k_{t2})\theta_{1x} - 2k_{t2}\theta_{2x} = 0 \quad J_2 \ddot{\theta}_{2x} + 2k_{t2}\theta_{2x} - 2k_{t2}\theta_{1x} = 0 \tag{5.80}$$

Equations (5.80) permit formulation of the mass matrix

$$[M] = \begin{bmatrix} J_1 & 0 \\ 0 & J_2 \end{bmatrix} \tag{5.81}$$

and of the stiffness matrix

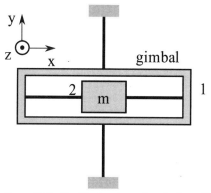

Figure 5.30   Double-torsion microdevice with cross-axes.

$$[K] = \begin{bmatrix} k_{t1} + k_{t2} & -k_{t2} \\ -k_{t2} & k_{t2} \end{bmatrix} \qquad (5.82)$$

The dynamic matrix $[A]$ is formulated by means of Eq. (5.53), and the corresponding resonant frequencies are

$$\omega_{1,2}^2 = \frac{J_1 k_{t1} + J_2(k_{t1} + k_{t2}) \pm \sqrt{[J_1 k_{t2} + J_2(k_{t1} + k_{t2})]^2 - 4 J_1 J_2 k_{t1} k_{t2}}}{J_1 J_2} \qquad (5.83)$$

Example: Study the resonant frequencies of the microsystem shown in Fig. 5.30.

The microdevice of Fig. 5.30 is similar to that discussed previously, but the one analyzed here has its hinge axes at 90°. This device, too, can be modeled as a 2-DOF system where the generalized coordinates are the rotation angles θ_x and θ_y. A more complete model would also look at the hinges bending and at the corresponding motions in a 4-DOF system. However, when only the torsion is of interest, as is the case here, the 2-DOF model is sufficiently accurate. The kinetic energy of the cross-axes torsional resonator of Fig. 5.30 is

$$T = \frac{1}{2} J_{1y} \dot{\theta}_y^2 + \frac{1}{2} J_{2x} \dot{\theta}_x^2 \qquad (5.84)$$

The potential energy of the same system is

$$U = k_{t1} \theta_y^2 + k_{t2} \theta_x^2 \qquad (5.85)$$

By applying again Lagrange's equations, the dynamic equations are obtained whose mass matrix is

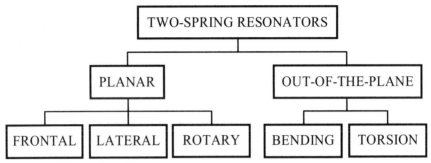

Figure 5.31 Two-spring microresonators.

$$[M] = \begin{bmatrix} J_{2x} & 0 \\ 0 & J_{1y} \end{bmatrix}$$ (5.86)

and stiffness matrix is

$$[K] = \begin{bmatrix} 2k_{t2} & 0 \\ 0 & 2k_{t1} \end{bmatrix}$$ (5.87)

This system is fully decoupled as both the mass matrix and the stiffness matrix are in diagonal form. The eigenvalues of the dynamic matrix which corresponds to the mass and stiffness matrices of Eqs. (5.86) and (5.87) yield the following resonant frequencies:

$$\omega_1 = 2\sqrt{\frac{k_{t1}}{J_{1y}}} \qquad \omega_2 = 2\sqrt{\frac{k_{t2}}{J_{2x}}}$$ (5.88)

5.3 Spring-Type Microresonators

Micro- and nanoresonators can be built with several complex elastic suspensions, including those based on beam-type springs, as previously discussed. The simplest multispring microresonators utilize two springs to achieve the resonant motion, as indicated in Fig. 5.31.

The planar designs are capable of motion in a plane parallel to the substrate, and Fig. 5.32 shows the sketches of frontal and lateral motion solutions. The x direction vibration corresponds to the frontal solution whereas the y direction solution is the lateral motion design.

In out-of-the-plane microresonators, the motion is no longer parallel to the substrate and can be produced by either bending or torsion; the corresponding lumped-parameter models are sketched in Fig. 5.33a and b.

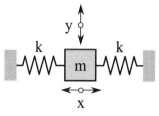

Figure 5.32 Two-spring planar microresonator.

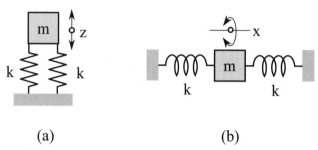

(a)   (b)

Figure 5.33 Two-spring out-of-the-plane microresonators: (a) bending; (b) torsional.

Several spring designs that are currently utilized in microresonators and enable parallel motion of a resonant device above the substrate are presented next. Figures. 5.34 through 5.38 are illustrations of spring pairs which realize both the suspension of the microresonator and the elasticity (compliance) function. More details regarding stiffness calculations for those springs are provided by Lobontiu and Garcia,[7] for instance. Figure 5.34 shows a proof mass which is symmetrically supported by two U springs. The actuation and sensing are performed electrostatically by means of comb-type units. This microsystem is designed to produce unidirectional frontal motion of the proof mass about the x direction.

Another planar elastic suspension is the folded beam, which is illustrated in Fig. 5.35 where the proof mass also vibrates about a single direction (the y direction). The designs of Figs. 5.34 and 5.35 are both of a frontal-type class.

Other two-spring planar microsuspensions are schematically shown in Figs. 5.36 through 5.38; they are solutions enabling planar translation of a proof mass about two perpendicular directions x and y. Eventually, full planar motion (with the z-direction rotation included) is also permitted by these spring designs.

The serpentine spring (shown in Fig. 5.36) is composed of several identical units that are serially connected. The units can also be scaled down or up in another design variant (see Lobontiu and Garcia[7] for instance),

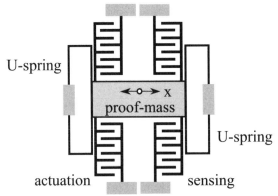

U-spring

proof-mass

U-spring

actuation   sensing

Figure 5.34   Proof mass with comb-type actuation and sensing and U spring suspension.

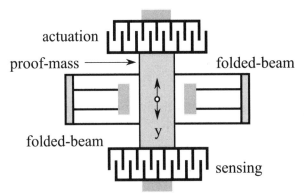

actuation

proof-mass ⟶   folded-beam

folded-beam y sensing

Figure 5.35   Proof mass with comb-type actuation and sensing and folded-beam suspension.

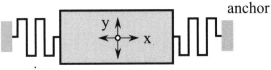

anchor

serpentine spring

Figure 5.36   Proof mass with serpentine microsuspension.

which is not shown here. The fishhook spring (more details on this design as well as on the relevant stiffnesses are given by Li et al.[8]) is another solution enabling two-direction planar motion as sketched in Fig. 5.37.

While the springs of Figs. 5.36 and 5.37 have different stiffnesses about the x and y motion directions due to their intrinsic asymmetry, the bent-beam serpentine spring—(Lobontiu and Garcia[7]) sketched in

fishhook spring

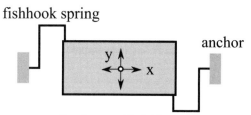

anchor

Figure 5.37 Proof mass with fishhook microsuspension.

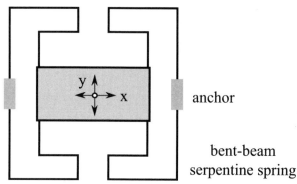

anchor

bent-beam
serpentine spring

Figure 5.38 Proof mass with bent-beam serpentine microsuspension.

Fig. 5.38—is symmetric and therefore is insensitive to the planar motion direction.

A mechanical resonator which utilizes specific springs enabling planar rotary motion is shown in Fig. 5.39, where both actuation and sensing are performed electrostatically by means of curved comb-type units. The spring microsuspension is a spiral which connects the mobile outer hub to a inner anchor for instance. Other solutions for rotary motion include straight or curved flexure hinges connecting the hub and the central anchor.

5.4 Transduction in Microresonators

In addition to the compliant structure, operation of microresonators is realized by means of actuation and/or sensing methods, which are together referred to as transduction. In actuation, a driving force or moment is applied such that the resonator is set into the desired vibrational state. Sensing is also needed, either stand-alone (as in pure sensors) or as the tool enabling evaluation of a resonator's state. Several means of transduction are available in MEMS/NEMS such as thermal, electrostatic, magnetic, electromagnetic, piezoelectric, piezomagnetic, optical, based on induced strain (bimorphs and multimorphs), with

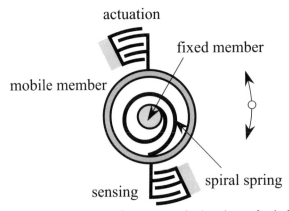

Figure 5.39 Rotary comb-type actuation/sensing and spiral-beam microsuspension.

shape memory alloys or fluid. Of all these methods, the electrostatic, electromagnetic, piezoelectric/piezomagnetic, and bimorph transduction methods are briefly discussed here, as these procedures, enable us to obtain excitation and detection frequencies which are within the range of mechanical microresonators.

5.4.1 Electrostatic transduction

The electrostatic actuation or sensing is one of the most popular techniques used with microresonators. Constructively, the electrostatic transduction can be implemented in planar and out-of-the-plane resonator designs. The comb-type electrostatic transduction is one of the most employed solutions for both planar and out-of-the-plane resonant applications. Electrostatic attraction forces can be generated between a fixed plate and a mobile one in different ways, depending on the boundary conditions pertaining to the mobile plate. The y motion shown in Fig. 5.40a can be activated, in the case where the mobile plate moves parallel to the fixed one, by keeping the gap constant. This type of transduction is known as *comb-finger*, and it basically consists of a planar longitudinal relative motion. The force generated through actuation is constant and can be expressed as

$$F_{cf} = \frac{\varepsilon l_z V^2}{2g} \tag{5.89}$$

where ε is the electrical permittivity, l_z is the plate overlap length about the z direction, V is the applied voltage, and g is the constant gap. Clearly, when a sinusoidal voltage is applied between the mobile and fixed plates, the resulting force will be sinusoidal as well, and a

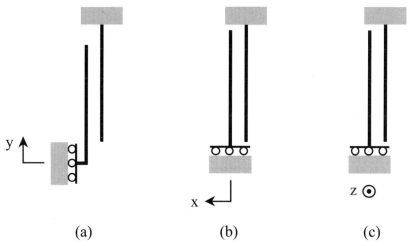

(a)   (b)   (c)

Figure 5.40 Electrostatic, comb-type transduction: (a) planar longitudinal; (b) planar transverse; (c) out-of-the-plane.

microresonator will operate in resonant conditions when the driving frequency and the mechanical system's natural frequency are identical. The capacity of the fixed-mobile plate couple varies when the mobile plate displaces by a quantity y in addition to the initial l_{0y} as

$$C_{cf} = \frac{\varepsilon(l_{0y} + y)l_z}{g} \qquad (5.90)$$

and this capacity variation can be transduced to a voltage variation in the sense circuit of a resonant microsensor, for instance.

Example: The microresonator sketched in Fig. 5.41a is supported by two beam springs. Both actuation and sensing are performed electrostatically by means of comb-type longitudinal units. Assume that damping is only produced by Couette-type losses due to the fluid-structure interaction taking place between the planar device and the substrate. Evaluate the average dynamic viscosity coefficient η.

The damping ratio can be expressed in terms of the Couette flow quality factor [Eq. (1.46)] according to Eq. (1.26) as

$$\xi = \frac{\omega z_0}{\pi \beta \mu \dot{x}^2} \qquad (5.91)$$

where z_0 is the fixed gap between the microdevice and the substrate, β is the actual-to-resonant frequency ratio [Eq. (1.16)], and μ is the dynamic viscosity.

The dynamic equation of motion is based on the lumped-parameter model of Fig. 5.41b, namely

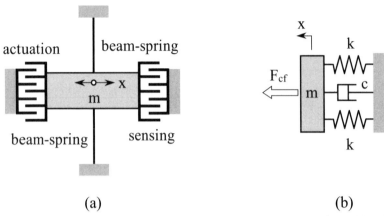

(a)   (b)

Figure 5.41 Electrostatic actuation and beam-spring suspension: (a) schematic microdevice; (b) lumped-parameter model.

$$m\,\ddot{x} + c\dot{x} + 2kx = F_0 \sin(\omega t) \qquad (5.92)$$

where it is assumed that the electrostatic force F_{cf} varies according to a sine law. By using Eqs. (5.91) and (1.13), Eq. (5.92) is reformulated as

$$\ddot{x} + \frac{\omega_r^2 z_0}{\pi\mu\dot{x}} + \omega_r^2 x = \frac{F_0}{m} \sin(\omega t) \qquad (5.93)$$

Equation (5.93) is nonlinear as the velocity is in the denominator of the second term of the left-hand side. This equation can be used to determine the average damping ratio by applying a numerical integration scheme, such as the Newmark procedure (see Wood,[9] for instance) which gives two equations connecting the displacement, velocity, and acceleration of a single-DOF system (such as the one studied here) at two consecutive moments in time in the form:

$$x_{i+1} = x_i + \Delta t\,\dot{x}_i + \frac{(\Delta t)^2}{2}(1 - \beta_2)\ddot{x}_i + \frac{(\Delta t)^2}{2}\beta_2\ddot{x}_{i+1}$$
$$\dot{x}_{i+1} = \dot{x}_i + \Delta t(1 - \beta_1)\ddot{x}_i + \Delta t\beta_1\ddot{x}_{i+1} \qquad (5.94)$$

In Eqs. (5.94), Δt is the sampling rate (the time interval between recording two consecutive capacitive measurements), whereas β_1 and β_2 are numerical control parameters, for values greater than 0.5, the Newmark scheme is unconditionally stable, as mentioned by Wood.[9] Equation (5.93) can also be sampled corresponding to the time stations i and $i + 1$, namely,

$$\ddot{x}_i = -\frac{\omega_r^2 z_0}{\pi\mu_i\dot{x}_i} - \omega_r^2 x_i + \frac{F_0}{m}\sin(\omega t_i)$$

$$\ddot{x}_{i+1} = -\frac{\omega_r^2 z_0}{\pi\mu_{i+1}\dot{x}_{i+1}} - \omega_r^2 x_{i+1} + \frac{F_0}{m}\sin(\omega t_{i+1})$$

(5.95)

Because the sensing is performed electrostatically, the microdevice displacement can be assessed at any time by Eq. (5.90) in terms of the measured capacitance, namely,

$$x_i = \frac{gC_i}{\varepsilon l_z} - l_{0x}$$

(5.96)

By combining Eqs. (5.94) and (5.95), the following two equations are produced:

$$\dot{x}_{i+1} = \dot{x}_i + \Delta t\,(1-\beta_1)\left[-\frac{\omega_r^2 z_0}{\pi\mu_i\dot{x}_i} - \omega_r^2 x_i + \frac{F_0}{m}\sin(\omega t_i)\right]$$

$$+\,\Delta t\,\beta_1\left[-\frac{\omega_r^2 z_0}{\pi\mu_{i+1}\dot{x}_{i+1}} - \omega_r^2 x_{i+1} + \frac{F_0}{m}\sin(\omega t_{i+1})\right]$$

(5.97)

$$x_{i+1} = x_i + \Delta t\,\dot{x}_i + \frac{(\Delta t)^2}{2}(1-\beta_2)\left[-\frac{\omega_r^2 z_0}{\pi\mu_i\dot{x}_i} - \omega_r^2 x_i + \frac{F_0}{m}\sin(\omega t_i)\right]$$

$$+\,\frac{(\Delta t)^2}{2}\beta_2\left[-\frac{\omega_r^2 z_0}{\pi\mu_{i+1}\dot{x}_{i+1}} - \omega_r^2 x_{i+1} + \frac{F_0}{m}\sin(\omega t_{i+1})\right]$$

By starting with zero-displacement and zero-velocity initial conditions, it is possible to determine the series η_i by working with Eqs. (5.97) in this sequence:

1. Determine the coefficient of dynamic viscosity η_{i+1} as a function of known displacements x_i and x_{i+1}, the velocity dx/dt at moment i, and the dynamic viscosity coefficient at the previous time moment η_i from the second of Eqs. (5.97).

2. Determine the velocity dx/dt at moment $i+1$ from the first of Eqs. (5.97).

The average coefficient of dynamic viscosity can eventually be determined as the arithmetic average of the n values of η_i.

The motion about the x direction can be enabled, as illustrated in Fig. 5.40b, in "parallel-plate" transduction where the mobile plate moves in-plane about a direction perpendicular to the plates. Out-of-the-plane motion is also possible when the mobile plate moves about the z axis due to fringe effects, either in a form of pure translation, as mentioned by Lee and Lin,[10] or as the result of a small relative rotation about a direction parallel to the x axis, as detailed by Mihailovich and MacDonald,[11] or by Selvakumar and Najafi.[2] To increase the transduction effects, several pairs of digits are utilized in comb-type microdevices. The force generated by plate-type attraction between two mating plates can be expressed as

$$F_p = \frac{\varepsilon A_s V^2}{2(g_0 + x)^2} \tag{5.98}$$

where A_s is the superimposed area, g_0 is the initial gap, and x is the distance traveled by the mobile plate, as indicated in Fig. 5.40b. It can be seen that, unlike the comb-finger actuation, the force in plate-type actuation varies nonlinearly with the displacement x. In sensing, the capacitance varies with the gap change x (see Lobontiu and Garcia,[7] for instance) as

$$C_p = \frac{\varepsilon A_s x}{(g_0 - x)^2} \tag{5.99}$$

and the sensitivity [the factor multiplying x in Eq. (5.99)] is not constant, as it was in comb-finger sensing.

Out-of-the-plane electrostatic transduction can also be achieved by means of microcantilevers and bridges whose bending motion can be transduced against a fixed plate. In either of the techniques mentioned here, the electrostatic variation between two plates can be generated by the source and control circuit of the device, and this leads to mechanical motion (in actuation), or the motion can be generated externally and sensed as a capacity variation by the corresponding circuit of the device (as in sensing).

5.4.2 Electromagnetic transduction

The principle of electromagnetic transduction in microresonators is based on the interaction between an external (usually constant) magnetic field and an alternating current (ac). The result is a Lorentz-type force which acts on the conductor carrying the current. Transverse vibrations could thus be generated in a fixed-fixed wire, as shown by Husain et al.,[12] for instance, according to the sketch of Fig. 5.42.

nanowire

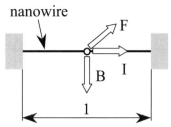

Figure 5.42   Nanowire for magnetomotive detection.

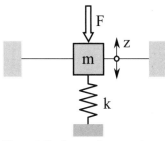

Figure 5.43   Lumped-parameter model of the nanowire under the action of the Lorentz force.

The Lorentz force which is generated through the interaction between the external magnetic field B and the current I is

$$\overline{F} = \overline{Il} \times \overline{B} \tag{5.100}$$

When the current is an ac input signal of the form:

$$I = I_0 \sin(\omega t) \tag{5.101}$$

and the external magnetic field is perpendicular to the wire, the sensed magnetomotive force is

$$F = I_0 l B \sin(\omega t) \tag{5.102}$$

Example: Determine the excitation frequency that will produce maximum vibrational amplitudes that are twice the static deflection for the case of a fixed-fixed constant circular cross-section wire by the interaction between a constant magnetic field B and an alternating current of amplitude I_0 and frequency ω passing through the wire.

Figure 5.43 shows the lumped-parameter model of the wire
The resonant frequency of the fixed-fixed wire is

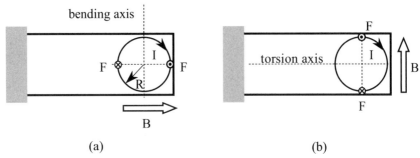

Figure 5.44 Cantilever-based electromagnetic sensing: (*a*) bending mode; (*b*) torsional mode.

$$\omega_r = 5.6 \frac{d}{l^2} \sqrt{\frac{E}{\rho}} \qquad (5.103)$$

When the forced vibrations are undamped, the amplitude ratio of Eq. (1.15) can be expressed as

$$\frac{X}{X_{st}} = \frac{1}{1 - (\omega / \omega_r)^2} \qquad (5.104)$$

For an amplitude ratio of 2, Eq. (5.104) in combination with Eq. (5.103) produces the excitation frequency

$$\omega = 3.955 \frac{d}{l^2} \sqrt{\frac{E}{\rho}} \qquad (5.105)$$

which is equal to approximately $0.71\omega_r$.

Another possibility of electromagnetic transduction is illustrated in Fig. 5.44 where a circular conducting wire is patterned to a micro-cantilever. When the external magnetic field is parallel to the micro-cantilever length, as in Fig. 5.44*a*, a couple is produced by the two opposite forces F which act at two diametrically opposed points, and the result is a bending moment which is applied about the axis shown in the same figure. In the case where the magnetic field is perpendicular to the cantilever length, as in Fig. 5.44*b*, the resulting Lorentz forces F will generate a couple acting along the member's longitudinal axis, and therefore the cantilever will undergo torsion. For any other direction of the magnetic field B situated in the microcantilever's plane, both bending and torsion will be produced. In both cases, the moment resulting through the interaction between the external magnetic field B and the current I yields a moment equal to

$$M = \pi R^2 BI \qquad (5.106)$$

Example: Determine the parameters defining the lumped-parameter undamped model of the microcantilever illustrated in Fig. 5.44a when the loop current is sinusoidal [Eq. (5.101)] and the center of the loop is placed at a distance l_1 from the free end of the microcantilever having the length l.

The parameters defining the lumped-parameter model of the single-DOF system of Fig. 5.1 are the mass m, stiffness k, and forcing amplitude F_0. The first two parameters are determined by Eqs. (2.66) and (2.61), respectively. To find the force which is located at the microcantilever's free end, the real case where a bending moment acts at a distance l_1 is made equivalent to the situation where a force acts at the free end, subject to the condition that both systems produce the same rotation at the point where the bending moment is applied. It can simply be shown that this requirement results in a force

$$F = \frac{2M}{l_1 + l} \qquad (5.107)$$

By combining now Eqs. (5.101), (5.106), and (5.107), the equivalent force can be expressed whose amplitude is

$$F_0 = \frac{2\pi R^2 B I_0}{l_1 + l} \qquad (5.108)$$

5.4.3 Piezoelectric and piezomagnetic transduction

Piezoelectric materials change dimension when exposed to a variation in an external electric field, whereas piezomagnetic materials change dimension when subjected to variations in an external magnetic field. In such situations microdevices that are designed based on this effect (also called reversed) behave as actuators. Conversely, when external mechanical pressure is applied to piezoelectric/piezomagnetic materials, they become electrically/magnetically polarized and therefore can be used as deformation sensors. Most often, piezoelectric materials (such as PZT, an alloy based on lead, zinc, and titanium; ZnO, zinc oxide; or $Al_{0.3}Ga_{0.7}As$, an alloy based on aluminum and gallium arsenide) and piezomagnetic materials (such as Terfenol-D) are deposited on substrate/structural layers and create sandwiched micro/nano cantilevers, bridges, or membranes. Piezomagnetic materials are sensitive to the relative position between the external magnetic field and their own polarization field. Positive magnetostrictive materials (see Jakubovics[13] for more details) do extend along the polarization direction when the two fields are parallel and contract about the polarization direction when the two fields are perpendicular.

compressed active layer

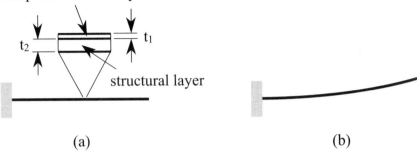

(a)   (b)

Figure 5.45 Bimorph with active piezoelectric layer in compression: (*a*) original (undeformed) state; (*b*) deformed state.

Several MEMS/NEMS transducers are designed as bimorphs that use piezoelectric or piezomagnetic materials. One component of the bimorph is the substrate (or structural layer) whereas the piezo (active) material is deposited over it. Variation of the electric/magnetic field leads to compression/extension of the active layer, but due to bonding to the substrate, the net result is upward/downward bowing of the entire sandwich structure. Figure 5.45, for instance, shows a bimorph which bends by prevented compression of an active piezoelectric layer.

An equivalent tip force F can be calculated which will produce the same tip rotation as the bending moment M which is created by the induced-strain actuation. It can simply be shown that the force is related to the moment in the form:

$$F = \frac{2M}{l} \qquad (5.109)$$

where l is the bimorph length. Lobontiu and Garcia[7] gave the relationship between the bending moment, the free induced strain ε_0, and the material/geometry properties of the bimorph, and therefore Eq. (5.109) can be reformulated as

$$F = \frac{4}{l} f \varepsilon_0 \qquad (5.110)$$

where $\quad f = \dfrac{E_1 E_2 A_1 A_2 (t_1 + t_2)(E_1 I_{y1} + E_2 I_{y2})}{E_1 A_2 \left\{ 4 E_1 I_1 + E_2 [4 I_{y2} + A_2 (t_1 + t_2)^2] \right\} + 4 E_2 A_2 (E_1 I_y + E_2 I_{y1} + E_2 I_{y2})} \qquad (5.111)$

For a piezoelectric material, the free strain is expressed in terms of the active layer thickness and applied voltage as

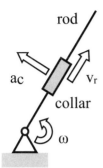

Figure 5.46   Collar sliding on a rotating rod as an example of the Coriolis effect.

$$\varepsilon_0 = d_{31}\frac{V}{t_1} \tag{5.112}$$

where d_{31} is the charge constant. When an ac voltage of the form:

$$V = V_0 \sin(\omega t) \tag{5.113}$$

is applied across the piezoelectric material, a sinusoidal force can be generated which is applied at the free tip, whose amplitude is

$$F_0 = \frac{4d_{31}}{lt_1}fV_0 \tag{5.114}$$

5.5   Resonant Microgyroscopes

Gyroscopes in the macroworld are bodies with a fixed position which can freely rotate about three independent axes under the action of external forces and moments. In the micro- and nanodomain, gyroscopes are mostly utilized as sensors of an externally applied angular velocity. Applications of microgyroscopes include automobile control, inertial navigation, platform stabilization, motion compensation in video cameras, inertial mouse devices in computers, virtual reality devices, robotics, and surgical instruments, as indicated by Fujita, Maenaka, and Maeda;[14] Li et al.;[15] Yang et al.;[16] Kawai et al.;[17] Geiger et al.;[18] Degani et al.;[4] Nakano, Toriyama, and Sugiyama;[19] Ajazi and Najafi;[20] Yazdi, Ayazi, and Najafi;[21] or Park et al.;[22] to cite just a few of the works dedicated to microgyroscopes.

The operating principle of gyroscopes is based on the Coriolis effect which is illustrated in Fig. 5.46.

The collar slides on the rod with a relative velocity v_r while the rod undergoes a rotary motion with an angular speed ω. It is known from

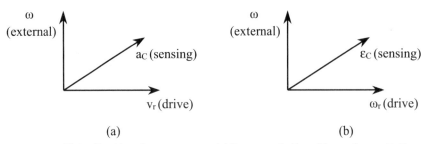

Figure 5.47   Main directions in a gyroscope: (*a*) linear excitation; (*b*) angular excitation.

dynamics that an additional acceleration is produced, which is named the Coriolis acceleration; its vector definition is

$$\bar{a}_C = 2\bar{\omega} \times \bar{v}r \tag{5.115}$$

and is directed as shown in Fig. 5.46. A better representation of the Coriolis acceleration is seen in Fig. 5.47*a* where the relative velocity and Coriolis acceleration are coplanar while the angular velocity direction is perpendicular to that plane.

The Coriolis effect is also produced when, instead of a linear relative velocity at the drive port, an angular relative velocity is used. On such occasions, the Coriolis effect produces an angular acceleration as shown in Fig. 5.47*b* which is calculated as

$$\bar{\varepsilon}_C = 2\bar{\omega} \times \bar{\omega}r \tag{5.116}$$

In MEMS/NEMS the relative velocity is produced through actuation, while the angular velocity is external. The combination of the two vectors results in the Coriolis acceleration which can be sensed about a direction perpendicular to the plane formed by the relative velocity and the Coriolis acceleration. The schematic representation of a micro-fabricated gyroscope which uses linear driving is illustrated in Fig. 5.48, where the outer rotating gimbal is assumed massless and the inner mass can translate about the local x and y axes.

An external angular velocity ω is applied to the entire gyroscope system, whereas a sinusoidal drive force is only applied to the vibrating mass about the drive direction. The combination between the external angular velocity ω and the relative motion of the mass about the x direction will produce a Coriolis acceleration about the sense (y) direction, as illustrated in Fig. 5.48. Assuming the relative velocity of the mass is directed about the positive x direction, the Coriolis acceleration will coincide with the positive y axis and will add to the

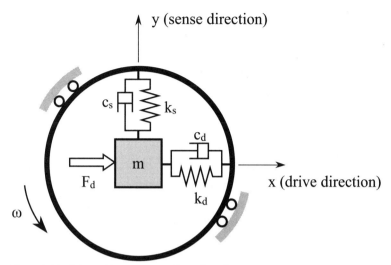

Figure 5.48 Schematic representation of a microgyroscope.

mass acceleration about that direction. The equations describing the motion of the 2-DOF gyroscopic system are

$$m\ddot{x} + c_d\dot{x} + k_d x = F_0 \sin(\omega_d t) \qquad m(\ddot{y} + 2\omega\dot{x}) + c_s\dot{y} + k_s y = 0 \quad (5.117)$$

where the subscript d indicates drive and the subscript s means sense. The first of Eqs. (5.117) can be solved independently for x, and the particular solution to it is of the form given in Chap. 1, namely,

$$x_p = X \sin(\omega_d t - \varphi_d) \qquad (5.118)$$

with $$X = \frac{F_0}{k_d\sqrt{(1 - \beta_d^2)^2 + (2\xi_d\beta_d)^2}} \qquad \varphi_d = \arctan\frac{2\xi_d\beta_d}{1 - \beta_d^2} \quad (5.119)$$

In Eqs. (5.119) the frequency and damping ratios are

$$\beta_d = \frac{\omega_d}{\omega_{d,r}} \qquad \xi_d = \frac{c_d}{2m\omega_{d,r}} \qquad (5.120)$$

where $\omega_{d,r}$ is the resonant frequency corresponding to the drive direction.

The second of Eqs. (5.117) can be rewritten as

$$\ddot{y} + 2\xi_s\omega_{s,r}\dot{y} + \omega_{s,r}^2 y = \frac{F_{0,y}}{m}\cos(\omega_d t - \varphi_d) \qquad (5.121)$$

where
$$F_{0,y} = -2m\omega\omega_d X \quad \xi_s = \frac{c_s}{2m\omega_{s,r}} \tag{5.122}$$

and $\omega_{s,r}$ is the resonant frequency of the sense branch of the gyroscope. The particular solution to this equation is

$$y_p = Y \sin(\omega_d t - \varphi_s) \tag{5.123}$$

The amplitude Y is

$$Y = \frac{F_{0,y}}{m\sqrt{(\omega_{s,r}^2 - \omega_d^2)^2 + (2\xi_s\omega_{s,r}\omega_d)^2}} \tag{5.124}$$

and the phase of the particular solution of Eq. (5.123) is

$$\varphi_s = 2\arctan\frac{2\xi_s\omega_{s,r}\omega_d}{\omega_{s,r}^2 - \omega_d^2 + (2\xi_s\omega_{s,r}\omega_d)^2 - [F_{0,y}/(mY)]^2}{\omega_{s,r}^2 - \omega_d^2 + F_{0,y}/(mY)} \tag{5.125}$$

The sense amplitude can be reformulated by using Eqs. (5.119) and (5.122) as

$$Y = \frac{2\omega\omega_d F_0}{m\omega_{d,r}^2\omega_{s,r}^2\sqrt{[(1-\beta_d^2)^2 + (2\xi_d\beta_d)^2][(1-\beta_s^2)^2 + (2\xi_s\beta_s)^2]}} \tag{5.126}$$

where
$$\beta_s = \frac{\omega_d}{\omega_{s,r}} \tag{5.127}$$

The following situations are possible:

1. Resonance in the drive branch: $\omega_d = \omega_{d,r}$. Because $\beta_d = 1$, the sense amplitude of Eq. (5.126) becomes

$$Y_d = \frac{\omega F_0}{\xi_d m\omega_{d,r}^2\omega_{s,r}^2\sqrt{(1-\beta_s^2)^2 + (2\xi_s\beta_s)^2}} \tag{5.128}$$

2. Resonance in the sense branch: $\omega_d = \omega_{s,r}$. In this situation $\beta_s = 1$, and therefore the sense amplitude of Eq. (5.126) transforms to

$$Y_s = \frac{\omega F_0}{\xi_s m\omega_{d,r}^2\omega_{s,r}\sqrt{(1-\beta_d^2)^2 + (2\xi_d\beta_d)^2}} \tag{5.129}$$

3. Resonance in both the drive and sense branches: $\omega_d = \omega_{d,r} = \omega_{s,r}$. in this particular case (the corresponding design is known as the well-tuned gyroscope), the drive and sense branches are identical in terms of both stiffness and damping, and $\beta_d = \beta_s = 1$. In addition, the driving frequency is equal to the resonant frequencies about the x and y directions. The sense amplitude of Eq. (5.126) becomes

$$Y_{ds} = \frac{\omega F_0}{2 \xi_d \xi_s m \omega_{d,r}^3} \qquad (5.130)$$

and definitely the sense amplitude is maximized.

As Eqs. (5.126), (5.128), (5.129), and (5.130) indicate, the external angular velocity ω can be determined in either of the four design cases in terms of the system's design parameters and assuming the sense displacement can be measured (which is most often performed by capacitive means in commercially available microfabricated gyroscopes).

Example: Compare the drive-resonance and sense-resonance sense amplitudes in the case where damping properties are identical for the drive and sense branches.

When the damping properties are identical for the drive and sense branches, Eqs. (5.128) and (5.129) can be combined into

$$\frac{Y_d}{Y_s} = \frac{\beta_s}{\beta_d} \sqrt{\frac{(1 - \beta_d^2)^2 + (2 \xi \beta_d)^2}{(1 - \beta_s^2)^2 + (2 \xi \beta_s)^2}} \qquad (5.131)$$

For a damping ratio of $\xi = 0.01$, Fig. 5.49 shows the three-dimensional plot corresponding to Eq. (5.131).

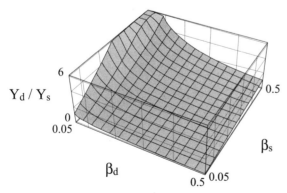

Figure 5.49 Sense amplitude ratio: drive resonance versus sense resonance – Eq. (5.131).

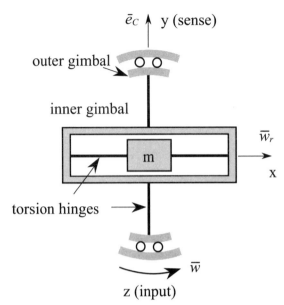

$\bar{e}_C$ ↑ y (sense)

outer gimbal ○ ○

inner gimbal

$\overline{w}_r$

m

x

torsion hinges ➤

○ ○

$\overline{w}$

z (input)

Figure 5.50 Two-gimbal microgyroscope with angular velocity driving.

As Fig. 5.49 suggests, the drive-resonance sense amplitude Y_d is larger than the sense-resonance sense amplitude Y_s for higher resonant drive frequencies (relatively small values of β_d) and smaller resonant sense frequencies (relatively large values of β_s). This relationship between the two amplitudes reverses for larger β_d and smaller β_s. It is therefore possible to obtain higher sense amplitudes by designing the drive and sense units such that their resonant frequencies are well separated.

The other driving option is to use angular relative velocity instead of linear velocity, as mentioned at the beginning of this section. An example of this principle is shown in the sketch of Fig. 5.50, which illustrates the solution proposed by the Charles Stark Drake Lab in the early 1990s.

The input motion consists of the angular velocity ω which is applied about the z axis (perpendicular to the plane of the figure), as was the case with the linear drive gyroscope. Unlike that design, the drive motion consists of an angular velocity ω_d which is applied to the inner gimbal and is facilitated by the two torsion hinges that are located on the x axis.

As shown in Eq. (5.116), the result is a Coriolis-type acceleration which is directed about the y axis (the sense axis) and which produces rotation of the outer gimbal about the y axis, this motion being enabled by the other two torsional hinges which are aligned with the y axis.

The dynamic equations of motion for the drive and sense directions are

$$J_x\ddot{\theta}_x + c_d\dot{\theta}_x + k_d\theta_x = M_0\sin(\omega_d t)$$
$$J_y(\ddot{\theta}_y + 2\omega\dot{\theta}_c) + c_s\dot{\theta}_y + k_s\theta_y = 0 \tag{5.132}$$

where
$$k_d = 2k_{1t} \qquad k_s = 2k_{2t} \tag{5.133}$$

with k_{1t} and k_{2t} being the torsional stiffnesses of the inner and outer hinges, respectively.

By following an approach similar to the one presented for the linear drive microgyroscope, it can be shown that the particular solution of the drive direction is of the form:

$$\theta_x = \Theta_x\sin(\omega_d t - \varphi_d) \tag{5.134}$$

with
$$\Theta_x = \frac{M_0}{k_d\sqrt{(1-\beta_d^2)^2 + (2\xi_d\beta_d)^2}} \qquad \varphi_d = \arctan\frac{2\xi_d\beta_d}{1-\beta_d^2} \tag{5.135}$$

and
$$\xi_d = \frac{c_d}{2J_x\omega_{d,r}} \tag{5.136}$$

Similarly, the particular solution corresponding to the sense direction is expressed as

$$\theta_y = \Theta_y\sin(\omega_d t - \varphi_s) \tag{5.137}$$

with
$$\Theta_y = \frac{M_{0,y}}{J_y\sqrt{(\omega_{s,r}^2 - \omega_d^2)^2 + (2\xi_s\omega_{s,r}\omega_d)^2}} \tag{5.138}$$

and
$$\varphi_s = 2\arctan\frac{2\xi_s\omega_{s,r}\omega_d + \sqrt{\dfrac{(\omega_{s,r}^2 - \omega_d^2)^2 + (2\xi_s\omega_{s,r}\omega_d)^2}{-[M_{0,y}/(J_s\Theta_y)]^2}}}{\omega_{s,r}^2 - \omega_d^2 + M_{0,y}/(J_s\Theta_y)} \tag{5.139}$$

In Eqs. (5.138) and (5.139),

$$M_{0,y} = -2J_y\,\omega\omega_d \tag{5.140}$$

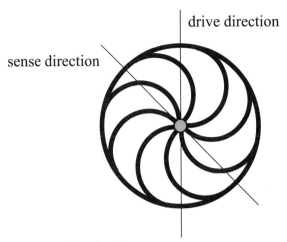

drive direction

sense direction

Figure 5.51   Vibrating ring gyroscope.

and
$$\xi_s = \frac{c_s}{2J_s\omega_{s,r}}$$
(5.141)

The particular cases of drive resonance ($\omega_d = \omega_{d,r}$), sense resonance ($\omega_d = \omega_{s,r}$), and full resonance ($\omega_d = \omega_{d,r} = \omega_{s,r}$) can be formulated from the generic Eq. (4.138). The well-tuned case ($\omega_d = \omega_{d,r} = \omega_{s,r}$), for instance, gives the following amplitude of the sense solution:

$$\Theta_{y,ds} = \frac{\omega M_0}{2\xi_d\xi_s J_x\omega_{d,r}^3}$$
(5.142)

It should be mentioned that squeeze-film damping dominates in this type of micro- and nanogyroscope, as opposed to slide-film damping, which was the prevalent damping mechanism in linear drive gyrosensors.

Another microfabricated gyroscope is the vibrating ring gyroscope (Ayazi and Najafi[20]) which consists of an elastic ring symmetrically supported by eight identical semicircular springs, as sketched in Fig. 5.51. This structure exhibits two flexural modes of equal resonant frequencies which are situated 45° apart, as sketched in Fig. 5.52. When a driving force is applied along the drive axis, the primary flexural mode is excited. Superposition of this motion to a rotary input about an axis perpendicular to the planar structure results in the secondary mode being excited, and this motion can be sensed capacitively.

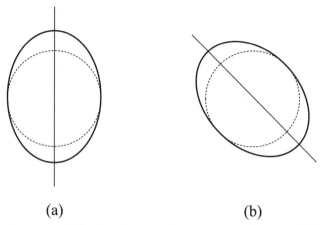

(a)   (b)

Figure 5.52   Modes in a vibratory ring gyroscope: (*a*) primary (drive) mode; (*b*) secondary (sense) mode.

5.6 Tuning Forks

Tuning forks have long been used as standard pitch tools for musical instrument calibration. Another utilization of tuning forks is in the clock and wristwatch industry as frequency standards. Resonating tuning forks, such as the ones introduced in Chap. 1, can also be employed as microsensors in a similar manner to gyroscopes in detecting changes in an external angular velocity, as shown by Satoh, Ohnishi, and Tomikawa;[23] Sato, Ono, and Tomikawa;[24] Momosaki;[25] or Matsiev.[26] The tuning forks operate similarly to gyroscopes, and the Coriolis effects are again the underlying principle. The classical tuning fork sensor is sketched in Fig. 5.53, where two variants are outlined. In the first variant, the driving is out-of-the-plane, which results in Coriolis accelerations pulling away the two tines in their plane (Fig. 5.53*a*). In the second variant (Fig. 5.53*b*), driving and sensing interchange, with the net result that the two tines are deformed out-of-the-plane through Coriolis effects.

The vibrational amounts involved in one tine of the tuning fork are sketched in Fig. 5.54. The input to the tine consists of the angular velocity ω which is applied about the z axis. A sinusoidal force is driving the tine about the x axis, and the result is a relative velocity v_r of the tip of the tine about the same axis. The interaction between the angular velocity and relative velocity results in a Coriolis-type acceleration about the y direction.

By ignoring the damping, the dynamic equations of motion about the drive (x) and sense (y) axes can be written, using a lumped-parameter model, as follows:

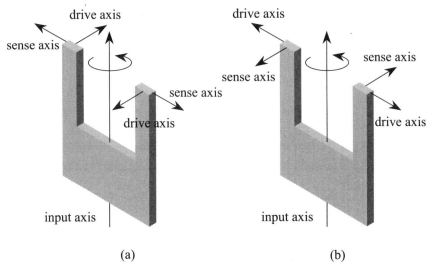

Figure 5.53　Classical tuning-fork microsensors: (*a*) out-of-the-plane driving; (*b*) in-plane driving.

$$m\ddot{x} + k_d x = F_0 \sin(\omega_d t) \quad m(\ddot{y} + 2\,\omega\dot{x}) + k_s y = 0 \qquad (5.143)$$

The particular solution to the first of Eqs. (5.143) is of the form:

$$x_p = X\sin(\omega_d t) \qquad (5.144)$$

where the amplitude is

$$X = \frac{F_0}{k_d(1 - \beta_d^2)} \qquad (5.145)$$

with β_d being the frequency ratio. Similarly, the particular solution of the second of Eqs. (5.143) is of the form:

$$y_p = Y\sin(\omega_d t) \qquad (5.146)$$

with
$$Y = -\frac{2\omega^2 F_0}{k_d k_s(1 - \beta_d^2)(1 - \beta_s^2)} \qquad (5.147)$$

When the sense amplitude Y is known, the input angular frequency can be determined from Eq. (5.147).

This model characterizing the dynamic response of a single tine can be utilized to evaluate the behavior of the two tines whose conjugate motion can be monitored electrostatically or by optical means.

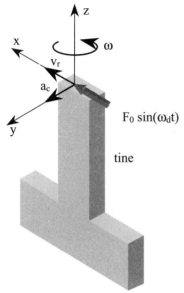

$F_0 \sin(\omega_d t)$

tine

Figure 5.54   Tine with Coriolis acceleration generated about the sense direction by rotation input and sinusoidal drive.

Similarly, a trident tuning fork, such as the one described by Satoh, Ohnishi, and Tomikawa,[23] for instance, can be monitored in terms of its Coriolis response by the same model. Figure 5.55 illustrates two possible utilizations of a trident tuning fork as a gyrosensor.

Another design is the double-ended tuning fork (DETF), which was introduced in Chap. 1. Figure 5.56 illustrates a double-ended tuning fork which is driven out of its plane in opposite directions. When an angular input is applied to the tuning fork about its long symmetry axes, Coriolis accelerations will act on both tines and stretch them apart. By reversing the driving forces, the tines will move deform inward and thus decrease their relative distance. In-the-plane driving (which is not shown here) is also possible, in the case where the Coriolis acceleration will generate out-of-the-plane deformations of the two tines. In terms of modeling the behavior of a double-ended tuning fork, it is sufficient to study one quarter model because of symmetry. The front view of a quarter model is sketched in Fig. 5.57.

The quarter-structure tuning fork sketched in Fig. 5.57a can be modeled as a fixed-guided beam, as illustrated in Fig. 5.57b, and therefore the model that has just been developed for an individual tine can be utilized here as well, with the mention that the lumped-parameter mass and inertia fractions need to be calculated for the

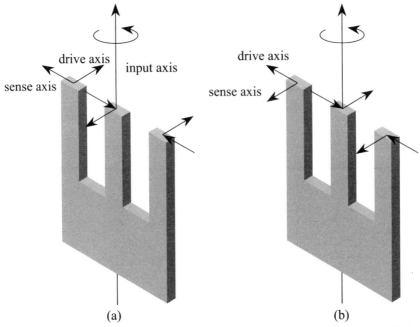

(a)   (b)

Figure 5.55   Trident tuning fork sensors: (*a*) out-of-the-plane driving; (*b*) in-plane driving.

fixed-guided boundary conditions, instead of the fixed-free boundary conditions which applied for the tine model.

Example: Determine the sense response of a double-ended tuning fork based on the fixed-guided beam quarter model. The length of a tine is $l = 100$ μm, and the cross section is rectangular with w parallel to the x direction in Fig. 5.57*b*. The material properties are $E = 160$ GPa, and $\rho = 2300$ kg/m^3, and the drive frequency ratio is $\beta_d = 0.7$.

It was shown in Chap. 4 that the lumped-parameter stiffnesses of a fixed-guided beam of length $l/2$ (half the tine, as shown in Fig. 5.57) are, according to Eq. (4.1),

$$k_d = \frac{96EI_y}{l^3} = \frac{8Ew^3t}{l^3} \qquad k_s = \frac{96EI_x}{l^3} = \frac{8Ewt^3}{l^3} \tag{5.148}$$

The lumped-parameter mass which needs to be placed at the guided end is, as indicated in Eq. (4.4),

$$m_b = \frac{13}{70}m \tag{5.149}$$

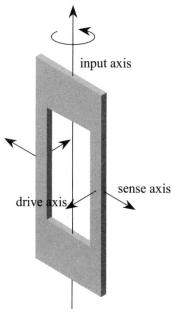

input axis

sense axis

drive axis

Figure 5.56 Double-ended tuning fork with out-of-the-plane driving and in-the-plane sensing.

where m is the mass of the entire tine (twice the length of the fixed-guided beam of Fig. 5.57b). The resonant frequencies about the drive and sense directions can be calculated based on Eq. (4.5) as

$$\omega_d = \sqrt{\frac{k_d}{m_b}} \qquad \omega_s = \sqrt{\frac{k_s}{m_b}} \qquad (5.150)$$

By using all these equations together with the numerical data of this example, it is possible to formulate the following function with the aid of Eq. (5.147):

$$f = \frac{Y}{2\omega^2 F_0} \qquad (5.151)$$

in terms of the nondimensional parameters

$$c_w = \frac{w}{l} \qquad c_t = \frac{t}{l} \qquad (5.152)$$

Figure 5.58 is the plot of f as a function of c_w and c_t. It can be seen that f, which is proportional to the sense amplitude Y, is higher for relatively smaller cross sections, as expected.

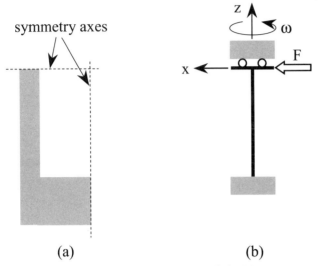

(a)   (b)

Figure 5.57 Model of double-ended tuning fork: (*a*) quarter structure; (*b*) fixed-guided beam model of quarter structure.

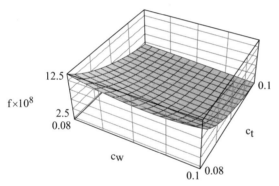

Figure 5.58 Plot of the *f* function, Eq. (5.152).

5.7   Resonant Accelerometers

There are two basic procedures to sense external acceleration by using resonant sensors. Both procedures are based on detecting shifts in a resonator's natural frequency as produced by changes in stiffness. The change in stiffness can be determined by an additional bending due to acceleration or to the action of an axial force, also produced by acceleration.

When the acceleration produces an additional bending moment on a microcantilever, for instance, which will displace the resonator's tip by a small quantity measured by the angle α, as shown in Fig. 5.59, the

Figure 5.59 Displaced fixed-free microcantilever.

flexure hinges

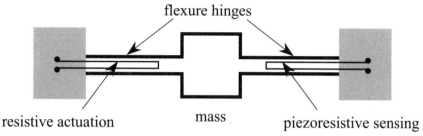

resistive actuation   mass   piezoresistive sensing

Figure 5.60 Paddle microbridge resonator.

bending stiffness is related to that of the original-position resonator as follows:

$$k_{b,e}^{*} = \frac{k_{b,e}}{\cos^2 \alpha} \qquad (5.153)$$

By assuming that the lumped-parameter mass remains the same, the modified resonant frequency is related to original one as

$$\omega_{b,e}^{*} = \frac{\omega_{b,e}}{\cos \alpha} \qquad (5.154)$$

which confirms that the resonant frequency of the inclined beam is larger than that of the original system. In other words, a slight bending from an external source will alter the resonant frequency.

One solution to sensing external acceleration through modification of the bending stiffness is sketched in Fig. 5.60, where a paddle microbridge is utilized as an acceleration microsensor, as proposed by Ohlckers et al.[27]

An acceleration which is applied about a direction perpendicular to the structure's plane will slightly displace the central mass, and thus the stiffness of the supporting flexure hinges will change, together with the bending resonant frequency of the entire structure. The actuation in this design was resistive and was provided in one hinge, whereas the sensing was piezoresistive and the corresponding circuit was diffused into the other flexure hinge.

flexure hinge

resonant beam

seismic mass

anchor

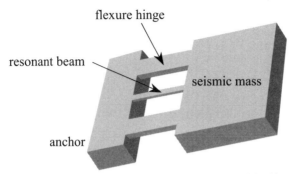

Figure 5.61   Resonant beam microaccelerometer with side supporting flexure hinges.

Another solution for sensing an external acceleration through bending-generated modification of the relevant resonant frequency is sketched in Fig. 5.61 and was utilized by Burrer, Esteve, and Lora-Tamayo[28] and Roszhart et al.,[29] for instance. The beam microresonator is located in the center and is flanked by two side flexure hinges supporting the seismic mass, which will move under the action of an external acceleration. This action will produce bending of the central beam, which will alter its resonant frequency.

Tudor et al.[30] propose utilizing a double-ended tuning fork instead of the simple beam resonator in Fig. 5.61. The particular solution with a double-ended tuning fork presents the advantage that the externally applied acceleration can be monitored by means of Coriolis effects (as shown in the previous section of this chapter) in addition to the normal change in the resonator's resonant frequency.

Huang et al.[31] proposed utilizing one single supporting flexure hinge and two side beam resonators, as sketched in Fig. 5.62. This specific design could be tailored to sense external acceleration which is applied either in plane (case where the seismic mass will move laterally) or out of the plane.

A similar design was mentioned by Esashi[32] and is sketched in Fig. 5.63. The seismic mass this time is supported by two side torsion hinges, whereas the beam resonator remains in a central position. Its clamping point to the seismic mass goes up or down out of the plane of the structure, and therefore the beam resonator bends upward or downward, which modifies the resonant frequency.

Burns et al.[33] microfabricated and tested a resonant microbeam accelerometer for sensing out-of-the-plane acceleration. Figure 5.64 sketches the top view of the accelerometer which comprises four resonant beams and has a seismic mass supported by flexure pairs at its corners.

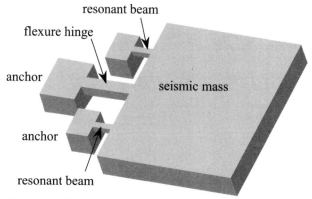

Figure 5.62 Resonant beam microaccelerometer with central supporting flexure hinge and two side beam resonators.

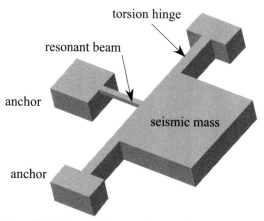

Figure 5.63 Resonant beam microaccelerometer with side supporting torsion hinges.

Another source of resonant frequency changing is the action of an axial force on a vibrating beam, as mentioned previously. It was shown in Chap. 2 how the bending resonant frequency varies when an axial force is applied to a fixed-free beam, and Fig. 5.65 shows the basic principle of this class of resonant accelerometers where the acceleration is applied axially to a beam resonator by means of a seismic mass usually.

Seshia et al.[34] developed a resonant microaccelerometer based on resonance alteration through acceleration-generated axial force, and Fig. 5.66 is a simplified sketch of that design.

The motion of the seismic mass which is produced by the external acceleration compresses one tuning fork while extending the other by equal amounts and therefore lowers and raises their resonant

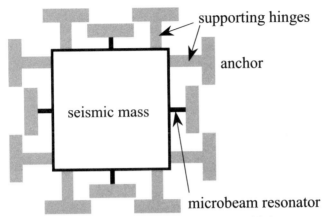

Figure 5.64 Resonator beam microaccelerometer with four resonators.

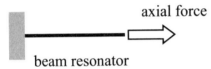

Figure 5.65 Cantilever beam resonator under the action of an acceleration-generated axial force.

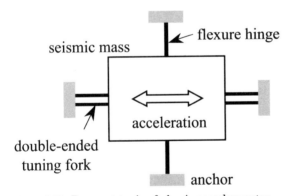

Figure 5.66 Resonant tuning fork microaccelerometer.

frequencies, respectively. Not shown in Fig. 5.66 is a leverage mechanism used to amplify the small mechanical motion of the seismic mass.

Another design has been proposed by Aikele et al.[35] where the mechanical microresonator is attached transversely to the seismic mass, as illustrated in Fig. 5.67, and therefore it is in either tension or compression, depending on the acceleration-generated motion of the seismic mass.

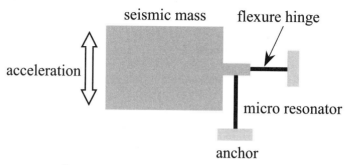

seismic mass   flexure hinge

acceleration

micro resonator

anchor

Figure 5.67   Resonant bar microaccelerometer.

It should be mentioned that resonant gyroscopes and accelerometers are characterized by performance parameters such as resolution, drift, zero-rate output, angle random walk, scale factor accuracy, or full-scale range (for more details, see Lefevre[36] or Yazdi, Ayazi, and Najafi[21]. These performance criteria enable the ranking of resonant micro-devices into rate-grade, tactical-grade, and inertial-grade categories in an ascending-quality order.

References

1. Z. Xiao, X. T. Wu, W. Peng, and K. R. Farmer, An angle-based design approach for rectangular electrostatic torsion actuators, *Journal of Microelectromechanical Systems*, 10(4), 2001, pp. 561–568.

2. A. Selvakumar, and K. Najafi, A high-sensitivity z-axis capacitive silicon microaccelerometer with a torsional suspension, *Journal of Microelectromechanical Systems*, 7(2), 1998, pp. 192–200.

3. F. P. Beer, and E. R. Johnston, Jr., *Vector Mechanics for Engineers—Dynamics,* 6th ed., McGraw-Hill, New York, 1996.

4. O. Degani, D. J. Sater, E. Socher, S. Kaldor, and Y. Nemirowski, Optimal design and noise generation of micromachined vibrating rate gyroscope with modulated integrative differential optical sensing, *Journal of Microelectromechanical Systems*, 7(3), 1998, pp. 329–338.

5. F. Ayela, and T. Fournier, An experimental study of anharmonic micromachined silicon resonators, *Measurement Science Technology*, **9**, 1998, pp. 1821–1830.

6. D. A. Harrington, P. Mohanty, and M. L. Roukes, Energy dissipation in suspended micromechanical resonators at low temperatures, *Physica B*, 284–288, 2000, pp. 2145–2146.

7. N. Lobontiu and E. Garcia, *Mechanics of Microelectromechanical Systems,* Kluwer Academic Press, New York, 2004.

8. Z. Li, Z. Yang, Z. Xiao, Y. Hao, G. Wu, and Y. Wang, A bulk micromachined lateral gyroscope fabricated with wafer bonding and deep trench etching, *Sensors and Actuators*, **83**, 2000, pp. 24–29.

9. W. L. Wood, *Practical Time Stepping Schemes*, Clarendon Press, Oxford, 1990.

10. K. B. Lee and L. Lin, Vertically supported microresonators, *12th International Conference on Solid State Sensors, Actuators and Microsystems*, 2003, pp. 847–850.

11. R. E. Mihailovich, and N. C. MacDonald, Dissipation measurements of vacuum-operated single-crystal silicon microresonators, *Sensors and Actuators A*, **50**, 1995, pp. 199–207.

12. A. Husain, J. Hone, H. W. Ch. Postma, X. M. H. Huang, T. Drake, M. Barbic, A. Scherer, and M. L. Roukes, Nanowire-based very-high-frequency electromechanical resonators, *Applied Physics Letters*, **83**(6), 2003, pp. 1240–1242.

13. J. P. Jakubovics, *Magnetism and Magnetic Materials*, University Press, Cambridge, England, 1994.

14. T. Fujita, K. Maenaka, and M. Maeda, Design of two-dimensional micromachined gyroscope by using nickel electroplating, *Sensors and Actuators A*, **66**, 1998, pp. 173–177.

15. Z. Li, Z. Yang, Z. Xiao, Y. Yao, T. Li, G. Wu, and Y. Wang, A bulk micromachined lateral gyroscope fabricated with wafer bonding and deep trench etching, *Sensors and Actuators A*, **83**, 2000, pp. 24–29.

16. H. Yang, M. Bao, H. Yin, and S. Shen, A novel bulk micromachined gyroscope based on a rectangular beam-mass structure, *Sensors and Actuators A*, **96**, 2002, pp. 145–151.

17. H. Kawai, K.-I Atsuchi, M. Tamura, and K. Ohwada, High-resolution microgyroscopes using vibratory motion adjustment technology, *Sensors and Actuators A*, **90**, 2001, pp. 153–159.

18. W. Geiger, B. Folkmer, U. Sobe, H. Sandmaier, and W. Lang, New designs of micromachined vibrating rate gyroscopes with decoupled oscillation modes, *Sensors and Actuators A*, **66**, 1998, pp. 118–124.

19. S. Nakano, T. Toriyama, and S. Sugiyama, Sensitivity analysis for a piezoresistive rotary movement micro gyroscope, *2001 International Symposium on Micromechanics and Human Science*, pp. 87–92.

20. F. Ayazi, and K. Najafi, A HARPSS polysilicon vibrating ring gyroscope, *Journal of Microelectromechanical Systems*, **10**(2), 2001, pp. 169–179.

21. N. Yazdi, F. Ayazi, and K. Najafi, Micromachined inertial sensors, *Proceedings of IEEE*, 1998, pp. 1640–1659.

22. K. Y. Park, C. W. Lee, Y. S. Oh, and Y. H. Cho, Laterally oscillated and forced-balanced vibratory rate gyroscope supported by fish hook shape springs, *Proceedings of IEEE*, 1997, pp. 494–499.

23. A. Satoh, K. Ohnishi, and Y. Tomikawa, Characteristics of the piezoelectric vibratory gyrosensor constructed using a trident tuning-fork resonator, *Proceedings of IEEE Ultrasonics Symposium*, **1**, 1998, pp. 555–558.

24. K. Sato, A. Ono, and Y. Tomikawa, Simulation of quartz crystal gyro-sensor using double-ended tuning fork resonator for detection of two-axial angular velocity, *2003 IEEE Symposium on Ultrasonics*, **2**, 2003, pp. 1350–1353.

25. E. Momosaki, A brief review of progress in quartz tuning fork resonators, *1997 IEEE International Frequency Control Symposium*, 1997, pp. 352–365.

26. L. F. Matsiev, Application of flexural mechanical resonator to high throughput liquid characterization, *Proceedings of the IEEE Ultrasonics Symposium*, **1**, 2000, pp. 427–434.

27. P. Ohlckers, R. Holm, H. Jakobsen, T. Kvisteroy, G. Kittisland, M. Nese, S. M. Nilsen, and A. Ferber, An integrated resonant accelerometer microsystem for automotive applications, *International Conference on Solid State Sensors and Actuators*, **2**, 1997, pp. 843–846.

28. C. Burrer, J. Esteve, and E. Lora-Tamayo, Resonant silicon accelerometers in bulk micromachining technology—an approach, *Journal of Microelectromechanical Systems*, **5**(2), 1996, pp. 122–130.

29. T. V. Roszhart, C. de Cotiis, H. Jerman, and J. Drake, An inertial-grade micromachined vibrating beam accelerometer, *IC Sensors Custom Product Papers and Briefs*, 1998, pp. 19–22.

30. M. J. Tudor, M. V. Andres, K. V. H. Foulds, and J. M. Naden, Silicon resonator sensors: interrogation techniques and characteristics, *IEE Proceedings on Control Theory and Applications*, **135**(5), 1988, pp. 364–368.

31. S. Huang, X. Li, Y. Wang, J. Jiao, X. Ge, D. Lu, L. Che, K. Zhang, and B. Xiong, A piezoresistive accelerometer with axially stressed tiny beams for both much increased sensitivity and much broadened bandwidth, *International Conference on Solid State Sensors, Actuation and Microsystems*, **1**, 2003, pp. 91–94.

32. M. Esashi, Resonant sensors by silicon micromachining, *Proceedings of the 1996 IEEE International Frequency Control Symposium*, 1996, pp. 609–614.

33. D. W. Burns, R. D. Horning, W. R. Herb, J. D. Zook, and H. Guckel, Resonant microbeam accelerometers, *The 8th International Conference on Solid State Sensors and Actuators*, **2**, 1995, pp. 659–662.

34. A. A. Seshia, M. Palaniapan, T. A. Roessig, R. T. Howe, R. W. Gooch, T. R. Schimert, and S. Montague, A vacuum packaged surface micromachined resonant accelerometer, *Journal of Microelectromechanical Systems*, **11**(6), 2002, pp. 784–795.

35. N. Aikele, K. Bauer, W. Ficker, F. Neubauer, U. Prechtel, J. Schalk, and H. Seidel, Resonant accelerometer with self-test, *Sensors and Actuators A*, **96**, 2002, pp. 145–151.

36. H. Lefevre, *The Fiber-Optic Gyroscope*, Artech House, Norwood, 1993.

6

Microcantilever and Microbridge Systems for Mass Detection

6.1 Introduction

The emerging field of micro and nano electromechanical (MEMS and NEMS) oscillators has fueled a renaissance in the field of resonant sensors and actuators with an unending flow of producing smaller and better electromechanical devices while providing a closely coupled link between the physical, chemical, and biological worlds. These systems have gained a wide theoretical interest and practical application in the field of sensors and actuators and have recently been adopted in valuable analytic instruments. In contrast to their macro counterparts such as quartz-crystal balances, surface-acoustic waves (SAWs), or flexural plates, they perform with increased functionality and complexity for various chemical and biological sensing applications. At the root of excitement in nanotechnology, compact electromechanical sensing devices offer a significant increase in analysis speed while suppressing the consumption of both samples and reagents. Furthermore, they can be used for a variety of different sensing purposes and offer unique possibilities by extending the dynamic range and ultimate sensitivity several orders of magnitude above those of their macro counterparts including conventional quartz-crystal oscillators. To perform their specialized functions, resonant sensors and actuators must reliably store and convert different forms of energy, transduce signals, and respond repeatably to external chemical and biological environments.

Generally, biomolecular adsorption of target analytes to functionalized regions of a cantilever-based sensor can alter mechanical stress within the oscillator and its total mass and thus influence both the bending and the natural frequency of the cantilever, respectively. Signal transduction is generally achieved by employing an optical deflection (or interferometric) system to measure the mechanical bending or the frequency spectra resulting from additional loading by the adsorbed mass. Within such a configuration, a collimated laser beam is focused onto the free end of the cantilever and is reflected onto a split photodiode. The dc offset of the difference signal between the two cells of the photodiode quantifies the cantilever bending while the resonant peak of the ac signal extracted by a spectrum analyzer corresponds to the natural frequency of the cantilever. In the case of interferometric detection, an ac change in the intensity of the reflected light corresponds to the cantilever natural frequency.

Most of the current work has been devoted to the immobilization of target species onto the surface of the resonating structure. In such a scenario, pathogen binding events are not confined to a particular portion of the device and can occur anywhere on the surface. Since both the resonant frequency shift and deflection are highly dependent on the position of the adsorbed material, it is difficult to determine the exact amount of additional mass present without any visual inspection. To circumvent these limitations, one can construct arrays of surface micromachined oscillators with precisely positioned catalyzing anchors. The incorporation of prefabricated adsorption sites allows adequate control of chemical surface functionality for the detection of analytes of interest. For example, by using electron beam lithography, stress-free polycrystalline silicon and low-stress silicon nitride micromachined resonators with evaporated gold contact pads can be fabricated. Alkanethiol molecules can be subsequently adsorbed from solution onto the Au anchors, creating a dense thiol monolayer with the tail end group pointing outward from the surface. A common feature of the alkanethiol self-assembled monolayer (SAM) systems is the strong interaction between the functional group and the gold substrate. The van der Waals interactions among the molecules permit dense packing of the monolayer into a supermolecular hierarchical organizations of interlocking components. Typically, the total amount of material in a well-packed alkanethiol SAM on gold is approximately 8.3×10^{-10} mol·cm^{-2}. Alkanethiolates offer unique opportunities for precise tailoring of the length of the alkane chain and chemical properties of both the head and tail groups, thus making them excellent systems for further engineering of the chemical surface functionality following the assembly of the SAM. Due to their extreme flexibility, SAMs allow the creation of ordered supermolecular structures, thereby making them attractive building blocks for superlattices and molecularly tailored surface properties.

Mass detection can be performed in the nano- and microdomain by means of relatively simple devices such as cantilever- or bridge-based systems whose experimentally monitored static deflection or resonant frequency shift offers quantitative assessment of the mass that attaches to such a device. Static deflection methods are structured around the fact that adsorbed matter induces a stress gradient into the structure, which produces deformations (deflections) in beam structures. On the other hand, mass that attaches in either a pointlike or a layerlike manner can be considered as a gravity force acting on a beam and generating deflections. Resonant detection methods rely on the change of the sensing system's mass (through attachment of extraneous substances) or on the combined alteration of mass and stiffness (as is the case with layerlike deposition) which produces a shift in the relevant resonant frequency (usually bending or torsional).

This chapter includes the study of mass detection by means of static deflection interrogation methods which analyze the gravity effects of point or layer forces in conjunction with the stiffness change (in the case of layer forces), but is mainly dedicated to studying evaluation methods of the resonant frequency shift.

The drive toward ever-smaller mechanical resonators is simple: minute amounts of deposited mass in the realm of femtograms and even attograms (10^{-18} g) have already been detected, and the promise of downscaling 3 orders of magnitude, which is equivalent to molecule-level detection, can only be achieved by very small-dimension devices which enable detection of significant resonant shifts.

Nano- and micromass detectors are therefore implemented as sensors in a variety of applications such as chemical, biological, or clinical analysis; environmental control; and monitoring of industrial applications by study of variations in temperature, viscosity, mass, stress, or electric/magnetic fields (Raiteri et al.[1]). Nanocantilevers, nanobridges, or systems based on these components are generally the structural implementations of these detectors. The original structure is coated with a layer that will capture the substance of interest by generally chemical reactions; this coating process, known as *functionalizing*, is followed by mass addition. Comparison between the significant amount values (deflection in static methods and frequency shift in resonant methods) measured before and after mass immobilization constitutes the metric for evaluating the quantity of attached mass. The static deflection method can be used, among other numerous applications, to detect hydrogen or mercury vapors by employing functionalized microcantilevers whose deflections can be as small as 1 nm (Britton et al.[2] or Baselt et al.[3]).

Resonant nano- and microcantilevers are the main tool of choice in conducting resonant mass detection, and the modeling approach comprises resonant frequency evaluation (of both the original system and the altered system), systems sensitivity, and evaluation of the deposited mass quantity (Dufour and Fadel[4]). Garcia et al.[5] proposed a shape optimization method for microcantilevers that can be utilized in mass detection and atomic force microscopy applications. Examples of mass detection through resonant methods are quite numerous, and the following sample inevitably mentions just a few. Brown et al.[6] presented the cantilever-in-cantilever microresonator which was utilized to monitor the levels of external (viscous) damping through changes in air pressure by means of magnetic actuation and piezo-resistive detection. Kawakatsu et al.[7] studied the design and performance of nanometric oscillators built as head-and-neck structures that can operate in the 0.5-GHz domain and can be implemented in scanning probe microscopy. Rogers et al.[8] used microcantilevers with integrated piezoelectric film for both actuation and sensing to detect mercury vapors that were adsorbed onto a gold layer with concentrations as low as 93 parts per billion (ppb). Pinnaduwage et al.[9] reported the use of silicon microcantilevers with gold surface and a self-assembled monolayer acid to capture the presence of plastic explosive substances in the range of 10 to 30 parts per trillion (ppt).

Ilic et al.[10] designed and studied the resonant response of an array of silicon nitride cantilevers that were capable of detecting the attachment of 16 cells (the equivalent of 6×10^{-12} g of mass) of *Eschericia coli* cells on antibody layer under ambient conditions. Similar cantilevers were reported by Ilic et al.[11] in applications that captured the presence of heat-killed *E. coli* cells attaching to a reactive *E. coli* antibody substance coated on the microcantilevers. The experiment enabled detection of a single cell (a mass of 665 fg was calculated from the resonant frequency shift, which was consistent with other cell mass evaluations). Sekaric et al.[12] studied the performance of thin-film crystalline-diamond nanoresonators that operated in the 640-MHz range due to the elevated sound velocity of the diamond and the very small cantilever dimensions (lengths in the micrometer range and thickness of 80 nm). Paddle nanostructures were utilized by Sekaric et al.[13] in an experimental design which used laser-light pumping to diminish the viscous damping losses in air and therefore to substantially improve the quality factors. Constant rectangular cross-section and paddle microcantilevers with thicknesses of 50 to 100 nm were analyzed by Lavrik and Datskos[14] to detect chemisorption of thiol molecules at the femtogram level by using

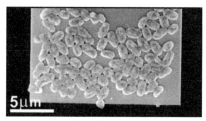 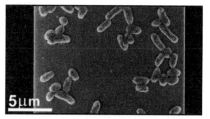

Figure 6.1 Scanning electron microscope (SEM) imaging of randomly dispersed molecules on microcantilevers. (*Courtesy of Dr. Ilic, the Cornell NanoScale Facility.*)

photothermal excitation and interferometric readout. Ilic, Yang, and Craighead[15] reported the use of paddle nanocantilever arrays that were coated with antibody agents in order to detect baculoviruses in solution with mass sensitivities of 10^{-19} g/Hz. The designs and associated experimental characterization enabled detection of the mass of a single virus which is 3×10^{-15} g approximately. Ilic et al.[16] designed and characterized paddle cantilever and bridge designs that were fabricated of polycrystalline silicon and silicon nitride. These structures were covered with gold dots as small as 50 nm in diameter, which enabled localized mass detection of a thiol monolayer.

An important performance metric factor of nanoresonators is the quality factor, which quantifies losses. Evoy et al.[17] analyzed the dissipation by temperature-dependent internal friction in paddle bridge nanoresonators operating in the megahertz range, by investigating the shifts in both the flexural and torsional resonant frequencies. Yasumara et al.[18] performed Q measurements that reached levels of 30,000 for 170-Å-thick, 5-μm-wide, and 80-μm-long microcantilevers. Yang, Ono, and Esashi[19] analyzed the losses and associated Q factors of water adsorption by means of 60 to 170-nm-thick cantilevers. Tamayo et al.[20] proposed a Q control method for liquid biosensing applications where magnetic excitation and photodetection were applied in experiments that were capable of recording quality factors 3 orders of magnitude higher than the regular ones.

Figures 6.1 through 6.7 are pictures of various resonant microdevices (one also shows the corresponding resonant response) that have been designed and experimentally tested for mass deposition detection by Dr. Rob Ilic and coworkers at the Cornell NanoScale Facility.

This chapter briefly analyzes the main traits of detecting mass deposition by using the static approach, but mainly focuses on the resonant shift method of detecting mass addition. Constant rectangular and variable-cross-section microcantilevers and microbridges are studied in conjunction with their abilities to capture the effects of added mass that can be immobilized in either pointlike or layerlike manner.

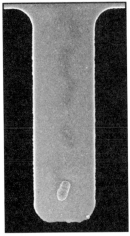

Figure 6.2   Single *E. coli* cell immobilized on a cantilever. (*Courtesy of Dr. Ilic, the Cornell NanoScale Facility.*)

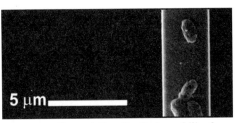

Figure 6.3   Several *E. coli* cells immobilized on a microcantilever. (*Courtesy of Dr. Ilic, the Cornell NanoScale Facility.*)

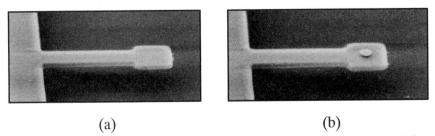

(a)   (b)

Figure 6.4   Paddle nanocantilever: (*a*) as-fabricated; (*b*) with 400-nm-diameter gold dot. (*Courtesy of Dr. Ilic, the Cornell NanoScale Facility.*)

The next section develops a general model for pointlike mass detection by resonant microdevices.

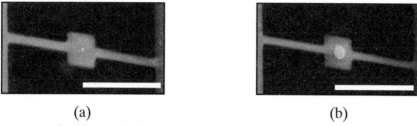

(a)   (b)

Figure 6.5 Paddle nanobridge with localized (*a*) 50-nm-diameter gold dot and (*b*) 400-nm-diameter gold dot. (*Courtesy of Dr. Ilic, the Cornell NanoScale Facility.*)

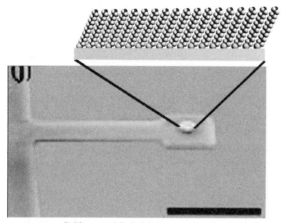

Figure 6.6   Self-assembled thiolate molecule deposited on a paddle cantilever gold dot. (*Courtesy of Dr. Ilic, the Cornell NanoScale Facility.*)

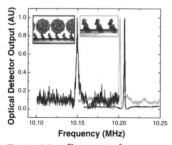

Figure 6.7   Resonant frequency shift of a 6-μm-long nanocantilever by capture of baculovirus. (*Courtesy of Dr. Ilic, the Cornell NanoScale Facility.*)

6.2   General Model of Point-Mass Addition Detection by Means of the Resonance Shift Method

When extraneous mass attaches in a pointlike manner on microdevices such as cantilevers or bridges, and resonant frequency methods are experimentally employed to determine the specific mass, only the quantity of deposited mass is of interest because the stiffness alteration is negligible in terms of modifying a relevant resonant frequency. The original resonant frequency of a certain micromember (be it cantilever, bridge, or membrane) can be expressed generically by means of lumped-parameter modeling as

$$\omega_{b,0} = \sqrt{\frac{k_{b,e}}{m_{b,e}}} \tag{6.1}$$

where $k_{b,e}$ and $m_{b,e}$ are the bending-related effective stiffness and mass, respectively. If torsion is monitored, Eq. (6.1) changes to

$$\omega_{t,0} = \sqrt{\frac{k_{t,e}}{J_{t,e}}} \tag{6.2}$$

where $k_{t,e}$ and $J_{t,e}$ are the torsion-related effective stiffness and mechanical moment of inertia. By considering the free vibrations as generic (and therefore they can be either bending- or torsion-generated), Eqs. (6.1) and (6.2) can be written in the single form:

$$\omega_0 = \sqrt{\frac{k_e}{m_e}} \tag{6.3}$$

Mass deposition on a micromember changes its original mass and therefore original resonant frequency such that the new resonant frequency becomes

$$\omega = \sqrt{\frac{k_e}{m_e + \Delta m}} \tag{6.4}$$

where Δm is the deposited mass. By combining Eqs. (6.3) and (6.4), the following relationship results between the original and modified resonant frequencies:

$$\frac{\omega_0}{\omega} = \sqrt{1 + f_m} \tag{6.5}$$

where

$$f_m = \frac{\Delta m}{m_e} \tag{6.6}$$

is the mass fraction. Equation (6.5) indicates that the resonant frequency decreases through mass addition, which means that

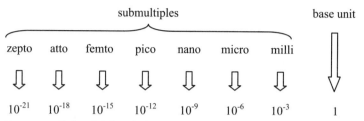

Figure 6.8   Submultiples of a base unit of measure and their mutual relationships.

$$\omega = \omega_0 - \Delta\omega \tag{6.7}$$

where $\Delta\omega$ is the resonant frequency variation. By analyzing Eqs. (6.5) and (6.7) it follows that an increase of the mass variation Δm leads to a corresponding increase in the resonant frequency variation $\Delta\omega$. Equations (6.5) and (6.7) enable us to define the mass sensitivity which represents the frequency variation per unit of added mass and which can be calculated as

$$\frac{\Delta\omega}{\Delta m} = \frac{\omega_0^2}{m(2\omega_0 + \Delta\omega)} \tag{6.8}$$

A mass deposition sensing device will possess high sensitivity when its resonant frequency is high and its mass is small, as Eq. (6.8) suggests.

It is also interesting to quantify the minimum mass quantity Δm_{min} which can be experimentally detected through a resonant frequency variation $\Delta\omega_{min}$. By using Eqs. (6.5) and (6.7), the following equation is obtained:

$$\Delta m_{min} = \frac{\Delta\omega_{min}(2\omega_0 + \Delta\omega_{min})}{\omega_0^2} m_e \tag{6.9}$$

This equation indicates that a decrease in the minimum detected mass can be achieved by a sensing device whose equivalent mass is small and whose resonant frequency is high, which, on aggregate, amounts to decreasing the mass of the vibrating micromember and increasing its stiffness. The drive toward miniaturization is therefore clear, as a smaller mass of the vibrating detector is possible only through size reduction, while keeping the relevant stiffness relatively large. Figure 6.8 illustrates the submultiples sequence for a generic unit of measure (which can be meter or gram, for instance).

Example: Study the amount of deposited mass that can be resonantly detected through bending by a constant rectangular cross-section

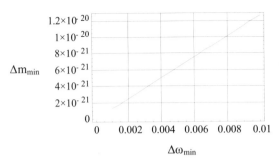

Figure 6.9   Minimum detected mass as a function of the minimum resonant frequency.

microcantilever constructed of polysilicon with $E = -170$ GPa and $\rho = 2330$ kg/m^3. The geometry of the microcantilever is defined by $l = 100$ μm, $w = 10$ μm, and $t = 50$ nm.

The mass of this micromember is $m = 1.165$ ng (nanograms), and its bending resonant frequency is calculated by means of Eq. (2.68) as $\omega_0 = 433{,}979$ rad/. Figure 6.9 plots the minimum added mass as a function of the minimum frequency variation.

By differentiating ω of Eq. (6.5) in terms of the mass addition Δm, the following equation is obtained:

$$\Delta\omega = -\frac{1}{2}\frac{k_e}{\sqrt{k_e/(m_e+\Delta m)}(m_e+\Delta m)^2}\Delta m \tag{6.10}$$

The minus sign in Eq. (6.10) indicates that an increase in the mass of the vibrating system by a quantity Δm will lead to a decrease in the resonant frequency by a quantity $\Delta\omega$. By ignoring the minus sign and by using the notation

$$f_\omega = \frac{\Delta\omega}{\omega_0} \tag{6.11}$$

Eq. (6.10) can be reformulated and written just in terms of the nondimensional parameters f_m and f_ω. It is thus possible to express the mass fraction as a function of the resonant frequency fraction in the form:

$$f_m = \frac{1 - 4f_\omega(1-f_\omega) - \sqrt{1 - 8f_\omega(1-f_\omega)}}{4f_\omega(1-f_\omega)} \tag{6.12}$$

Conversely, the same equation enables us to express f_ω in terms of f_m, namely,

$$f_\omega = \frac{1 + f_m + \sqrt{1 + f_m(4+f_m)}}{2(1+f_m)} \tag{6.13}$$

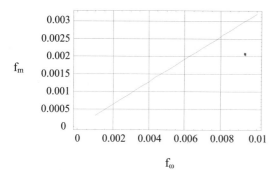

Figure 6.10 Mass fraction in terms of the bending resonant frequency fraction for a constant rectangular cross-section microbridge.

Example: Evaluate the mass fraction in terms of the resonant frequency fraction [Eq. (6.12)] as well as the resonant frequency fraction in terms of the mass fraction [Eq. (6.13)] by analyzing the free bending vibrations of a constant rectangular cross-section microbridge having the material and geometric properties given in the previous example.

The effective mass of this particular microbridge is 2.10×10^{-13} kg (0.21 ng), and the bending resonant frequency is 2.8 MHz. By using Eq. (6.12), the mass fraction is plotted against the bending resonant frequency in Fig. 6.10, and the converse relationship of Eq. (6.13) can be pictured in a figure which, for the same range of the variable, is identical to Fig. 6.10 and therefore is not plotted here.

As Fig. 6.10 indicates, a change of 0.001 in f_m results in an approximate change in f_ω of about 0.003, which in absolute figures means that a mass addition of 0.001×0.21 ng = 0.21 pg can be realized through a change in the resonant frequency of 2.8 MHz $\times$ 0.003 = 8.4 kHz.

6.3 Mass Detection by Means of Microcantilevers

The presence of extraneous substances can be detected by means of microcantilevers either statically (quasi-statically) when changes in deflections and/or rotations are experimentally monitored or dynamically when changes in the bending resonant frequency are, again, experimentally determined after the monitored substances deposited (attached) on a microcantilever by either physical means (adsorption, for instance) or chemical reactions. Two situations are investigated in this chapter, pertaining to either constant- or variable-cross-section microcantilevers. The case, where the deposited mass can be considered as a point mass, is analyzed first, whereas the second possibility, where the deposited extraneous substance is a layer, is studied afterward. As mentioned, experimental deformation or modal data are necessary in

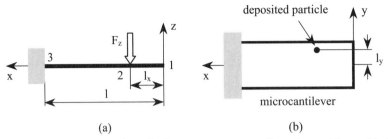

(a)   (b)

Figure 6.11   Point-mass detection by constant-cross-section microcantilever: (*a*) side view; (*b*) top view.

either case to evaluate the deposited mass, as two approaches can be pursued, namely, the static (quasi-static) procedure and the modal (resonant) one.

6.3.1 Constant-cross-section microcantilevers

Constant (usually rectangular) cross-section microcantilevers are constructively simple devices that enable mass deposition detection. This section details the mass detection process by considering first that the deposition occurs in a pointlike manner and then in a layer. For each situation, a static approach is presented as well as a resonance-based procedure.

Point-mass detection.   Mass can deposit (attach) locally on a very small area of the microcantilever, which may well be considered a point. When the microcantilever is monitored quasi-statically, the deposited mass might bend the member, and this deflection change can be determined experimentally. In case the experiment is conducted modally, the bending resonant response is altered by mass deposition, and the resulting resonant frequency shift enables quantification of the deposited mass. Both approaches are studied in the following subsections.

Static approach.   The problem of static detection of mass presence is briefly discussed next. A more detailed presentation of this subject is given by Lobontiu and Garcia.[21] A pointlike mass is first assumed that attaches to a microcantilever, as shown in Fig. 6.11. In essence, the mass quantity Δm, as well as its position on the microcantilever (quantified by l_x and l_y), can be determined by considering the deformations produced under the action of the deposited mass gravitational effects through bending and torsion at the free end, which are u_{1z}, θ_{1y}, and θ_{1x}, and which can be measured experimentally.

By using the notation of Fig. 6.11, the following static equations can be written:

$$u_{1z} = \frac{F_z(l - l_x)^2(2l + l_x)}{6EI_y}$$

$$\theta_{1y} = \frac{F_z(l - l_x)^2}{2EI_y}$$

$$\theta_{1x} = \frac{F_z l_y(l - l_x)}{GI_t}$$

(6.14)

By solving the equation system above, the following solution is obtained:

$$F_z = \frac{2EI_y\theta_{1y}^3}{9(l\theta_{1y} - u_{1z})^2}$$

$$l_x = \frac{3u_{1z}}{\theta_{1y}} - 2l$$

$$l_y = \frac{3GI_t\theta_{1x}(l\theta_{1y} - u_{1z})}{2EI_y\theta_{1y}^2}$$

(6.15)

When the detection system is set up to capture gravitational effects, the force F_z is

$$F_z = \Delta m g$$

(6.16)

and therefore the added mass can be determined from the first of Eqs. (6.15).

Resonant approach. Mass addition to a microcantilever modifies its resonant response, and it has been mentioned that the bending resonant frequency is the smallest one in an ascending series also containing the torsional and axial resonant frequencies. Figure 6.12 shows schematically a pointlike mass which is attached to a microcantilever at a distance a measured from the free end.

It was shown in Chap. 5 that the exact bending resonant frequency of a microcantilever is

$$\omega_{b,0} = 3.52\sqrt{\frac{EI_y}{ml^3}}$$

(6.17)

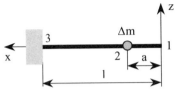

Figure 6.12   Pointlike mass deposited on a microcantilever.

where m is the total mass of the microcantilever. Because the bending stiffness at the free end of the microcantilever shown in Fig. 6.12 is

$$k_b = \frac{3EI_y}{l^3} \qquad (6.18)$$

and the effective mass which is located at the same point is

$$m_b = \frac{33}{140} m \qquad (6.19)$$

the bending resonant frequency in the presence of a deposited mass Δm can be expressed as

$$\omega_b = \sqrt{\frac{k_b}{m_b + \Delta m_e}} \qquad (6.20)$$

where Δm_e is the efficient deposited mass. It should be remembered that all calculations pertaining to lumped-parameter modeling use the free endpoint of a cantilever to locate both stiffness and mass. This is the reason why an efficient (or equivalent) deposited mass, denoted by Δm_e, which needs to be located at the free end, has to be calculated as it corresponds to the real additional mass Δm, which attaches at a distance a, as illustrated in Fig. 6.12. By applying Rayleigh's principle, which has been utilized several times thus far in this book, it can be shown that the effective deposited mass is

$$\Delta m_e = f_b(a)^2 \Delta m \qquad (6.21)$$

where $\qquad f_b(a) = 1 - \frac{3}{2}\frac{a}{l} + \frac{1}{2}\frac{a^3}{l^3} \qquad (6.22)$

is the bending distribution function introduced in Eq. (2.63) which, in the variant of Eq. (6.22), is the ratio of the beam deflection (or velocity) at point 2 to the deflection (or velocity) at point 1 in Fig. 6.12.

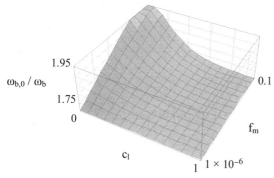

Figure 6.13 Frequency ratio in terms of length and mass fractions.

Equations (6.17) through (6.22) enable us to express the ratio of the original resonant frequency to the one that has changed through the modification in the system's mass, namely,

$$\frac{\omega_{b,0}}{\omega_b} = 2.03\sqrt{\frac{33}{140} + (1 - \frac{3}{2}\frac{a}{l} + \frac{1}{2}\frac{a^3}{l^3})^2 \frac{\Delta m}{m_b}} \qquad (6.23)$$

By using the substitutions

$$f_m = \frac{\Delta m}{m_b} \qquad c_l = \frac{a}{l} \qquad (6.24)$$

Equation (6.23) can be reformulated as

$$\frac{\omega_{b,0}}{\omega_b} = 0.297\sqrt{33 + 35(1 - c_l)^4(2 + c_l)^2 f_m} \qquad (6.25)$$

Figure 6.13 is the three-dimensional plot illustrating the variation of the frequency ratio defined and formulated in Eq. (6.25) as a function of the two nondimensional parameters of Eqs. (6.24).

The natural tendency, illustrated in Fig. 6.13, of the resonant frequency ratio is to decrease when the mass fraction increases and when the length ratio (fraction) decreases. A frequency ratio increase actually means a decrease in the modified frequency which is produced through an increase in the deposited mass (which means an increase of f_m) and/or a diminishing in a, the distance where the additional mass attaches to the microcantilever (which means a decrease in c_l).

Equation (6.25) can be rearranged in the form:

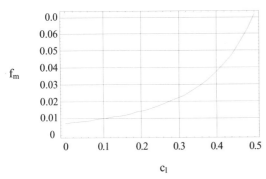

Figure 6.14   Mass fraction in terms of length fraction.

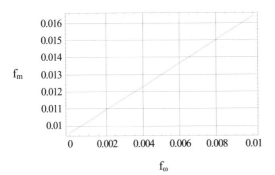

Figure 6.15   Mass fraction in terms of resonant frequency ratio.

$$f_m = \frac{33}{140}\frac{1.03/(1 - f_\omega)^2 - 1}{(1 - 1.5c_l + 0.5c_l^3)^2} \tag{6.26}$$

where the resonant frequency ratio is defined in Eq. (6.11). Figure 6.14 plots the mass fraction of Eq. (6.26) as a function of c_l for a frequency ratio $f_\omega = 0.0001$ whereas Fig. 6.15 plots the same function in terms of the frequency ratio for a value of $c_l = 0.1$.

Figure 6.14 indicates that as the length ratio increases (the attached mass gets closer to the root of the microcantilever), more added mass is needed to obtain the same frequency ratio of 0.0001. Similarly, Fig. 6.15 suggests that for a fixed position of the attached mass of the microcantilever, more attached mass (meaning a larger mass fraction f_m) will produce a larger resonant frequency shift (larger frequency ratio f_ω).

Layer-mass detection.   As mentioned previously, mass can attach to a microcantilever in the form of a layer, and experimental evaluation of static changes in the elastic deformation, as well as altering of the first

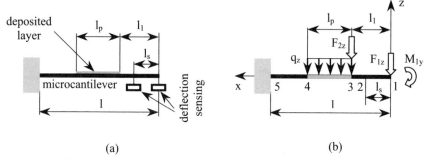

Figure 6.16 Layer-mass detection by constant rectangular cross-section microcantilever: (a) side view of setup; (b) side view of microcantilever with equivalent distributed load q_z and dummy loading F_{1z}, M_{1y}, F_{2z}.

(bending) resonant frequency, can be utilized to evaluate the quantity of the deposited mass, as shown in the next two subsections.

Static approach. Extraneous mass can attach over a length l_p on a constant-cross-section microcantilever, as sketched in Fig. 6.16. Given that deposition is produced over the entire width w (not shown in Fig. 6.16), and the length of the deposited layer is l_p, and this layer is positioned at l_s from the free end of the microcantilever, as shown in Fig. 6.16a, to determine the quantity of deposited mass Δm requires also knowledge of the lengths l_l and l_p; in other words, three experimental quantities are necessary. There can be two deflections—u_{1z} at the free end and u_{2z} measured at a distance l_s from the free end—as well as the slope (rotation) θ_{1y} at the same free end. These quantities can be measured quasi-statically by means of laser interferometry, for instance.

The situation pictured in Fig. 6.16a is equivalent to the model of Fig. 6.16b where the gravitational action of the deposited layer is substituted by a uniformly distributed load q_z acting over the length l_p. Over this length, the bending stiffness will also be modified from the original one, because the attached mass adds to the microcantilever. To express the three deformations, u_{1z}, θ_{1y}, and u_{2z}, three dummy loads F_{1z}, M_{1y}, and F_{2z} (all equal to zero in the end) are applied at points 1 and 2, respectively, also indicated in Fig. 6.16b. By applying Castigliano's displacement method, the three deformations are expressed as

$$u_{1z} = \frac{1}{E_1 I_{y1}} \left(\int_0^{l_s} M_1 \frac{\partial M_1}{\partial F_{1z}} dx + \int_{l_s}^{l_1} M_2 \frac{\partial M_2}{\partial F_{1z}} dx \right.$$

$$\left. + \int_{l_1+l_p}^{l} M_4 \frac{\partial M_4}{\partial F_{1z}} dx \right) + \frac{1}{(EI_y)_e} \int_{l_1}^{l_1+l_p} M_3 \frac{\partial M_3}{\partial F_{1z}} dx$$

$$\theta_{1y} = \frac{1}{E_1 I_{y1}} \left(\int_0^{l_s} M_1 \frac{\partial M_1}{\partial M_{1y}} dx + \int_{l_s}^{l_1} M_2 \frac{\partial M_2}{\partial M_{1y}} dx \right.$$

$$\left. + \int_{l_1+l_p}^{l} M_4 \frac{\partial M_4}{\partial M_{1y}} dx \right) + \frac{1}{(EI_y)_e} \int_{l_1}^{l_1+l_p} M_3 \frac{\partial M_3}{\partial M_{1y}} dx$$

$$u_{2z} = \frac{1}{E_1 I_{y1}} \left(\int_{l_s}^{l_1} M_2 \frac{\partial M_2}{\partial F_{2z}} dx + \int_{l_1+l_p}^{l} M_4 \frac{\partial M_4}{\partial F_{2z}} dx \right)$$

$$+ \frac{1}{(EI_y)_e} \int_{l_1}^{l_1+l_p} M_3 \frac{\partial M_3}{\partial F_{2z}} dx$$

(6.27)

The equivalent rigidity is calculated over the length l_p where the attached mass and microcantilever do superimpose and, as shown in a previous chapter, is equal to

$$(EI_y)_e = E_1[I_{y1} + z_1 A_1(z_1 - z_N)] + E_p[I_{yp} + z_p A_p(z_p - z_N)] \qquad (6.28)$$

The bending moments M_1 through M_4, which enter Eqs. (6.27), are

$$M_1 = -M_{1y} - F_{1z}x$$
$$M_2 = -M_{1y} - F_{1z}x - F_{2z}(x - l_s)$$
$$M_3 = -M_{1y} - F_{1z}x - F_{2z}(x - l_s) - \frac{q_z(x - l_1)^2}{2} \qquad (6.29)$$
$$M_4 = -M_{1y} - F_{1z}x - F_{2z}(x - l_s) - q_z l_p \left[x - \left(\frac{l_1 + l_p}{2} \right) \right]$$

and the corresponding partial derivatives of the same Eqs. (6.27) can simply be calculated from Eqs. (6.29).

The subscript 1 denotes the microcantilever (substrate for the dissimilar-length sandwich microcantilever), and the subscript p indicates the patch (deposited mass). The cross-sectional properties of the two components are:

$$A_1 = wt_1 \quad A_p = wt_p \quad I_{y1} = \frac{wt_1^3}{12} \quad I_{yp} = \frac{wt_p^3}{12} \tag{6.30}$$

The coordinates z_1 and z_p are

$$z_1 = \frac{t_1}{2} \quad z_p = t_1 + \frac{t_p}{2} \tag{6.31}$$

whereas the position of the neutral axis is calculated, also previously shown, as

$$z_N = \frac{z_1 E_1 A_1 + z_p E_p A_p}{E_1 A_1 + E_p A_p} \tag{6.32}$$

By considering that the uniformly distributed load q_z (which is defined as a force per unit length) is gravity-generated as

$$q_z = \frac{\Delta m g}{l_p} \tag{6.33}$$

Eqs. (6.27) can be written into the following form after we perform partial derivation and integration:

$$u_{1z} = \frac{l_p \left\{ l_p^2 (4l_1 + 3l_p) \big/ (EI_y)e + 2[2(l - l_1)^2 (2l + l_1) - 3l_p (l^2 - l_1^2) - l_p^3 \big|](E_1 I_{y1})] \right\} q_z}{24}$$

$$\theta_{1y} = \frac{l_p \, [\, l_p^2 \big| (EI_y)_e + 3(l - l_1)(l - l_1 - l_p) \big/ (E_1 I_{y1})] \, q_z}{6} \tag{6.34}$$

$$u_{2z} = \frac{l_p \left\{ l_p^2 (4l_1 + 3l_p - 4l_s) \big/ (EI_y)e + 2(l - l_1 - l_p)[\, 4l^2 - 2l_1^2 - l_1 l_p + l_p^2 + l(l_p - 2l_1 - 6l_s) + 6l_1 l_s \,] \big/ (E_1 I_{y1})] \right\} q_z}{24}$$

This equation system involves solving of higher-degree algebraic equations, which can be done numerically for specified geometry and material parameters.

In the case where the microcantilever is functionalized over a length l_p (which is known a priori) and when this functionalized patch (which is assumed to be covered entirely by the attached mass) starts from the microcantilever's free end ($l_1 = 0$), the deposited mass can be calculated as either

$$\Delta m = \frac{24u_{1z}}{g[3l_p^3/(EI_y)_e + 2(4l^3 - 3l^2l_p - l_p^3)/(E_1I_{y1})]} \tag{6.35}$$

or

$$\Delta m = \frac{6\theta_{1y}}{g[l_p^2/(EI_y)_e + 3l(l - l_p)/(E_1I_{y1})]} \tag{6.36}$$

depending on whether the tip deflection u_{1z} [as shown in Eq. (6.35)] or the tip slope θ_{1y} [as in Eq. (6.36)] is available experimentally.

Resonant approach. Finding the attached mass by the resonant method implies measuring the shift in the bending resonant frequency after mass attachment, as a result of alterations in both the stiffness and the mass of the microcantilever-based system. Again, after the extraneous substance attaches to the original microcantilever, the compound system behaves as a dissimilar-length sandwich, whose equivalent lumped-parameter stiffness and inertia were discussed in Chap. 3. Figure 3.35, which has been utilized for dissimilar-length bimorph cantilever calculations, is also valid in the present case. By taking into consideration the expressions determined for the equivalent stiffness $k_{b,e}$ and the effective mass $m_{b,e}$ [Eqs. (3.172) and (3.174), respectively], the added mass Δm (which is identical to the patch mass in the dissimilar-length sandwich microcantilever model) is determined as

$$\Delta m = \frac{l_p}{l_p'}\left(\frac{140}{33}\frac{k_{b,e}}{\omega^2} - m_1\right) \tag{6.37}$$

where m_1 is the original microcantilever mass and ω is the altered bending resonant frequency. Equation (6.37) is particularly useful when the modified resonant frequency is available experimentally.

The bending frequency ratio can be expressed as

$$\frac{\omega_{b,0}}{\omega_b} = \sqrt{\frac{k_{b,0}}{k_b}}\sqrt{\frac{m_b}{m_{b,0}}} \tag{6.38}$$

where the subscript 0 indicates the original condition (no mass deposited on the microcantilever). By using the equations corresponding to the original stiffness and to the modified one [Eqs. (2.61) and (3.172), respectively] the stiffness ratio of Eq. (6.38) can be expressed as

$$\frac{k_{b,0}}{k_b} = 1 - \frac{c_t c_E c_{lp}(3c_{l1}^2 + 3c_{l1}c_{lp} + c_{lp}^2)\{3 + c_t[6 + c_t(4 + c_E c_t)]\}}{1 + c_E c_t\{4 + c_t[6 + c_t(4 + c_E c_t)]\}} \quad (6.39)$$

where the nondimensional parameters are defined as

$$c_t = \frac{t_p}{t_1} \qquad c_{lp} = \frac{l_p}{l} \qquad c_{l1} = \frac{l_1}{l} \qquad c_E = \frac{E_p}{E_1} \qquad (6.40)$$

Similarly, the mass ratio of Eq. (6.38) is expressed by means of Eqs. (2.66), (3.174), (3.175), and (3.176) as

$$\frac{m_b}{m_{b,0}} = 1 + \frac{c_\rho c_t c_{lp}}{33}\left[1 + \sum_{\substack{i=1 \\ i\neq 5}}^{6} c_i \frac{(c_{l1} + c_{lp})^{i+1} - c_{l1}^{i+1}}{c_{lp}}\right] \quad (6.41)$$

with the coefficients c_i being $c_1 = -210$, $c_2 = 105$, $c_3 = 35$, $c_4 = -42$, $c_6 = 5$, and

$$c_\rho = \frac{\rho_\rho}{\rho_1} \qquad (6.42)$$

Example: Study the influence of the nondimensional parameters of Eqs. (6.40) and (6.42) on the bending resonant frequency ratio in the case of layerlike mass deposition on a cantilever whose parameters are ranging as $c_{l1} \rightarrow [0, 0.9]$, $c_{lp} \rightarrow [0, 1]$, $c_t \rightarrow [0.01, 1]$, $c_E \rightarrow [0.1, 1]$, $c_\rho \rightarrow [0.5, 5]$.

By considering that $c_{l1} = 0$, $c_E = 0.2$, and $c_\rho = 0.5$, the plot of Fig. 6.17 shows the variation of the resonant frequency ratio in terms of c_{lp} and c_t. Figure 6.17 indicates that the patch length parameter c_{lp} has a notable influence on the bending resonant frequency ratio in cases where the patch thickness compares to the substrate thickness, when this ratio reaches a maximum. The frequency ratio increases with the thickness ratio except for the extreme cases where the patch length is almost equal to the substrate length. Figure 6.18 is the three-dimensional plot of the same frequency ratio in terms of nondimensional material parameters for $c_{l1} = 0$, $c_{lp} = 0.2$, and $c_t = 0.2$. It can be seen that variation by a factor of 10 in the longitudinal elasticity modulus leaves the resonant frequency almost unchanged. Increasing the density of the patch relatively to the density of the substrate increases the frequency ratio.

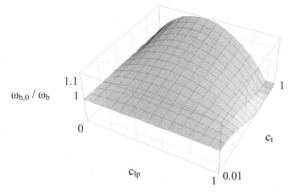

Figure 6.17 Bending resonant frequency ratio in terms of patch length and thickness nondimensional parameters.

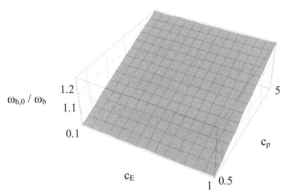

Figure 6.18 Bending resonant frequency ratio in terms of elasticity modulus and density nondimensional parameters.

Figure 6.19 plots the resonant frequency ratio as a function of the patch position nondimensional parameter c_{l1}. It has been plotted for $c_{lp} = 0.2$, $c_t = 0$, $c_E = 0.2$, and $c_\rho = 0.5$. While the original resonant frequency is generally higher than the altered one, for cases where the patch is situated toward the fixed root of the microcantilever, the relationship between the two resonant frequencies reverses, as a relatively small effective mass will result from a real small patch mass, which will increase the resonant frequency of the microcantilever.

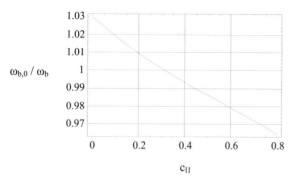

Figure 6.19 Bending resonant frequency ratio in terms of patch position nondimensional parameter.

6.3.2 Variable-cross-section microcantilevers

Variable-cross-section microcantilevers can also be utilized for mass deposition detection in a manner similar to the one presented for constant-cross-section microcantilevers. Mass detection is treated here similarly to mass detection by means of constant-cross-section microcantilevers; therefore pointlike and layerlike mass detection is analyzed by using both the static approach and the resonant one.

Point-mass detection

Static approach. As a reminder, when a point mass attaches to a microcantilever, its position (as defined by the parameters l_x and l_y in Fig. 3.37) and the mass quantity Δm are the unknowns. By experimental evaluation of three elastic deformations at the free tip, for instance, u_{1z}, θ_{1y}, θ_{1x}, the three equations that were valid for a constant-cross-section configuration can be reformulated as

$$u_{1z} = \Delta m g (C_{2,l} - l_x C_{2,c}) \qquad \theta_{1y} = \Delta m g (C_{2,c} - l_x C_{2,r})$$
$$\theta_{1x} = \Delta m g l_y C_{2,t} \tag{6.43}$$

where the newly introduced compliances are calculated with respect to point $\dot{2}$ in Fig. 3.37a (which explains the first subscript) and are defined as

$$C_{2,r} = \frac{1}{E} \int_{l_x}^{l} \frac{dx}{I_y(x)} \quad C_{2,c} = \frac{1}{G} \int_{l_x}^{l} \frac{x\,dx}{I_y(x)}$$

$$C_{2,l} = \frac{1}{E} \int_{l_x}^{l} \frac{x^2 dx}{I_y(x)} \quad C_{2,t} = \frac{1}{G} \int_{l_x}^{l} \frac{dx}{I_t(x)}$$

(6.44)

The equation system (6.43), which includes the definition Eqs. (6.44), is nonlinear as the unknown l_x also intervenes in compliances, and therefore its solution implies using numerical techniques.

Resonant approach. In using the resonant approach to determine the quantity of pointlike deposited mass, the algorithm that has been detailed for constant-cross-section microcantilevers can be extended to variable-cross-section members. The effective mass which results after mass deposition is

$$m_b = m_{b,0} + f_b(a)^2 \Delta m \tag{6.45}$$

where $f_b(a)$ is the distribution function corresponding to the position where the mass Δm has deposited. This distribution function is given in Eq. (2.63) for a constant-cross-section microcantilever, and it was shown to also be an accurate approximation for variable-cross-section members. Equation (6.45) expresses the modified resonant frequency in the form:

$$\omega_b = \sqrt{\frac{k_{b,0}}{m_{b,0} + \Delta m\, f_b^2(a)}} \tag{6.46}$$

The direct use of Eq. (6.46) is to determine the amount of mass that locally deposits on a variable-cross-section microcantilever in terms of the altered bending resonant frequency (which can be determined experimentally), namely,

$$\Delta m = \frac{k_{b,0}/\omega_b^2 - m_{b,0}}{f_b^2(a)} \tag{6.47}$$

By also taking into account the original resonant frequency of a microcantilever, the following frequency ratio can be formulated:

$$\frac{\omega_{b,0}}{\omega_b} = \sqrt{1 + \left(1 - \frac{3}{2}c_l + \frac{1}{2}c_l^3\right)^2 f_m} \tag{6.48}$$

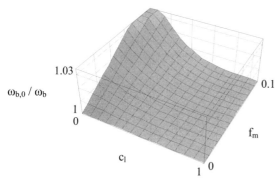

$\omega_{b,0} / \omega_b$

1.03

1

0

0.1

f_m

c_l

1   0

Figure 6.20 Bending resonant frequency ratio in terms of nondimensional length ratio and mass fraction.

where c_l and f_m are length and mass fractions, respectively, and have been defined in Eqs. (6.24). Equation (6.48) indicates that the altered resonant frequency is always lower than the original (the ratio is greater than 1). Another interesting peculiarity displayed by the resonant frequency ratio of Eq. (6.48) is that it does not depend on the particular properties (geometric and material) of a specific variable-cross-section microcantilever. Figure 6.20 is the three-dimensional plot of this frequency ratio as a function of the two nondimensional parameters. As the particle attaches farther away from the free tip, the resonant frequency ratio diminishes because the altered frequency increases relatively due to a decrease in the effective mass. Conversely, an increase in the mass fraction will produce a corresponding increase in the resonant frequency ratio.

The added mass can also be expressed in the following form:

$$\Delta m = \frac{1/(1 - f_\omega)^2 - 1}{(1 - 1.5c_l + 0.5c_l^3)^2} m \qquad (6.49)$$

where f_ω is the resonant frequency fraction and was defined in Eq. (6.11) in terms of the frequency shift and the original resonant frequency. Conversely, the resonant frequency shift is

$$\Delta \omega = \left[1 - \frac{1}{\sqrt{1 + (1 - 1.5c_l + 0.5c_l^3)^2 f_m}} \right] \omega_{b,0} \qquad (6.50)$$

Example: Compare the resonant frequency shift of a constant rectangular cross-section microcantilever of dimensions l, w, and t to that of a rectangular cross-section paddle microcantilever, as pictured in Fig. 6.21. Consider that $w_2 = 4w_1$, $w_1 = w$, and $l_1 + l_2 = l$, $l_1 = 80 \ \mu m$, $l_2 = 10 \ \mu m$, $t = 0.4 \ \mu m$,

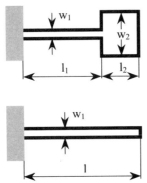

Figure 6.21 Top view of paddle and constant-cross-section microcantilevers.

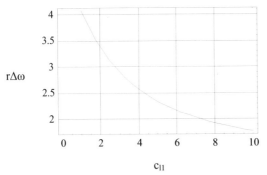

Figure 6.22 Constant-cross-section versus paddle microcantilevers: frequency ratio comparison in terms of the length nondimensional ratio.

$\rho = 2300$ kg/m^3, and $c_1 = 0.2$. The two microcantilevers are built of the same material and have identical thicknesses. The same point mass deposits in an identical localized manner on the two microcantilevers.

Figure 6.22 is the two-dimensional plot of the following frequency shift ratio:

$$r \, \Delta\omega = \frac{\Delta\omega}{\Delta\omega_{\text{paddle}}} \tag{6.51}$$

where $\Delta\omega$ is the frequency shift of the constant-cross-section microcantilever and $\Delta\omega_{\text{paddle}}$ is the similar shift of the paddle microcantilever. Both frequency shifts have been calculated by means of Eq. (6.50) via the corresponding equations defining the resonant frequencies and effective masses. The non-dimensional amount c_{l1} is the ratio of lengths l_2 to l_1. Figure 6.22 shows that the frequency ratio decreases with the parameter c_{l1} increasing, which indicates that the two microcantilever configurations will produce frequency shifts that are approximately equal when the length l_2 increases and l_1 decreases, which means that the paddle microcantilever approaches the constant-cross-section shape.

Layer-mass detection. When mass deposits as a thin layer over a variable-cross-section microcantilever, the change in the bending resonant frequency can be used as the quantifier indicator for calculation of the deposited mass as well as of its position along the microcantilever. Only the static approach is detailed here, as the resonant approach is identical to the one that applies to constant-cross-section microcantilevers, and which has been dealt with in this section.

Figure 3.39 is used again as a supporting schematic for layerlike mass deposition. In the case where the microcantilever has a variable cross section and after we express the bending moments and their corresponding derivatives, Eqs. (3.196) can be written in the form:

$$u_{1z} = \left[\frac{l_1^2}{2} C_{3,c} - l_1 C_{3,l} + \frac{1}{2} C_{3,d} - l_p \left(l_1 + \frac{l_p}{2} \right) C_{4,c} + l_p C_{4,l} \right] q_z$$

$$\theta_{1y} = \left[\frac{l_1^2}{2} C_{3,r} - l_1 C_{3,c} + \frac{1}{2} C_{3,l} - l_p \left(l_1 + \frac{l_p}{2} \right) C_{4,r} + l_p C_{4,c} \right] q_z$$

$$u_{2z} = \left[\frac{-l_s l_p^2}{2} C_{3,r} + l_1 \left(l_s + \frac{l_1}{2} \right) C_{3,c} - \left(l_1 + l_s \right) C_{3,l} + \frac{1}{2} C_{3,d} \right.$$

$$\left. + l_s l_p \left(l_1 + \frac{l_p}{2} \right) C_{4,r} - l_p \left(l_1 + l_s + \frac{l_p}{2} \right) C_{4,c} + l_p C_{4,l} \right] q_z \tag{6.52}$$

where the compliances are calculated as

$$C_{3,r} = \int_{l_1}^{l_1+l_p} \frac{dx}{(EI_y)_e} \quad C_{3,c} = \int_{l_1}^{l_1+l_p} \frac{x\,dx}{(EI_y)_e} \quad C_{3,l} = \int_{l_1}^{l_1+l_p} \frac{x^2\,dx}{(EI_y)_e}$$

$$C_{3,d} = \int_{l_1}^{l_1+l_p} \frac{x^3\,dx}{(EI_y)_e} \quad C_{4,r} = \int_{l_1+l_p}^{l} \frac{dx}{(EI_y)_e} \quad C_{4,c} = \int_{l_1+l_p}^{l} \frac{x\,dx}{(EI_y)_e} \tag{6.53}$$

$$C_{4,l} = \int_{l_1+l_p}^{l} \frac{x^2\,dx}{(EI_y)_e}$$

6.4   Mass Detection by Means of Microbridges

Mass capture can also be realized by means of microbridges. Utilization of microbridges in mass deposition detection is motivated in cases where higher resonant frequencies are expected than those produced by constructively similar or identical microcantilevers. Constant- and variable-cross-section bridge configurations are investigated in relationship to their performance in static or resonant mass detection.

6.4.1 Constant-cross-section microbridges

The first part of this subsection analyzes the static and resonant responses of constant-cross-section microbridges when extraneous mass attaches in either point- or layerlike fashion.

Point-mass detection. When the substance which is mobilized on a microbridge has small dimensions which warrant the pointlike mode to be employed, the stiffness of the original sensing microdevice remains unaltered, and the variations are either the deflected shape (in the static approach) or the mass, which causes modification of the resonant frequency (in the resonant approach), as discussed in the following.

Static approach. Figure 6.11b, which shows a constant-cross-section microcantilever where a mass attaches, can also be utilized here to find the quantity of deposited mass as well as its position on a half-length microbridge (in other words, the originally free end of the microcantilever is guided this time, and the length is only $l/2$). Three experimental measurements are needed again to determine Δm, l_x, and l_y. While measuring the deflection u_{1z} and torsional rotation θ_{1x} at the guided end is feasible, another amount needs to be determined, except for θ_{1y} which is zero in this case. Another deflection, for instance, can be monitored such as that at a point situated at an abscissa a past the location where the mass is deposited. The following equations are valid in this case:

$$u_{1z} = \frac{(l - 2l_x)^2(l + 4l_x)}{96EI_y}\Delta mg \quad u_{3z} = \frac{(l - 2a)^2(l + 4a)}{96EI_y}\Delta mg$$

$$\theta_{1x} = \frac{(l - 2l_x)l_y}{2GI_t}\Delta mg \tag{6.54}$$

The second of Eqs. (6.54) enables us to determine the quantity of deposited mass as

$$\Delta m = \frac{96EI_y u_{3z}}{(l - 2a)^2(l + 4a)g} \tag{6.55}$$

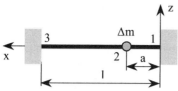

Figure 6.23   Pointlike mass deposited on a microbridge.

The first of Eqs. (6.54) is a third-degree equation which provides l_x, and the third of Eqs. (6.54) gives the value of l_y, provided Δm and l_x were determined, namely,

$$l_y = \frac{2GI_t}{(l - 2l_x)\Delta mg} \tag{6.56}$$

Resonant approach. A point mass which attaches to a microbridge of length l is schematically illustrated in Fig. 6.23.

The bending resonant frequency of a microbridge was shown to be

$$\omega_{b,0} = 22.373\sqrt{\frac{EI_y}{ml^3}} \tag{6.57}$$

where m is the total mass of the microbridge. The bending stiffness at the microbridge midpoint is

$$k_b = \frac{192EI_y}{l^3} \tag{6.58}$$

and the effective mass which is located at the same point is

$$m_b = \frac{128}{315}m \tag{6.59}$$

The bending resonant frequency in the presence of a deposited mass Δm is expressed in Eq. (6.20) that applied for a microcantilever where Δm_e is the efficient deposited mass, whose significance was explained in the subsection treating mass deposition through resonance by means of microcantilevers. For a microbridge, the efficient (or equivalent) deposited mass Δm_e is located at the midpoint and is again calculated by applying Rayleigh's principle according to Eq. (6.21) where

$$f_b(a) = 16\frac{a^2}{l^2}\left(1 - \frac{a}{l}\right)^2 \tag{6.60}$$

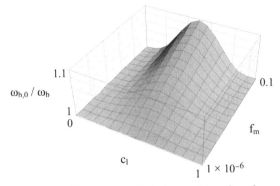

Figure 6.24   Frequency ratio in terms of length and mass fractions.

is the bending distribution function, introduced in Eq. (4.11), which is the ratio of the beam deflection (or velocity) at point 2 to the deflection (or velocity) at the midpoint in Fig. 6.23.

Similarly to the algorithm developed for microcantilevers, the ratio of the original microbridge resonant frequency to the altered resonant frequency is expressed as

$$\frac{\omega_{b,0}}{\omega_b} = 1.569\sqrt{\frac{128}{315} + \frac{256a^4(l-a)^4}{l^8}\frac{\Delta m}{m_b}} \tag{6.61}$$

By using the substitutions introduced in Eqs. (6.24), Eq. (6.61) is reformulated as

$$\frac{\omega_{b,0}}{\omega_b} = 1.569\sqrt{0.406 + 256c_l^4(1-c_l)^4 f_m} \tag{6.62}$$

The frequency ratio of Eq. (6.62) is plotted in Fig. 6.24 against the length and mass fractions.

Figure 6.24 indicates that the resonant frequency ratio increases when the mass fraction increases. And the explanation that has been provided for the similar trend displayed by microcantilevers is also valid, namely, that increasing the frequency ratio actually means reducing the resonant frequency, which is produced by the increase in the deposited mass (which means an increase of f_m). Figure 6.24 also illustrates the symmetric dependency of the frequency ratio on the mass location. The maximum resonant frequency shift is registered at the microbridge midpoint because the inertia effect of the deposited mass at that point is maximum.

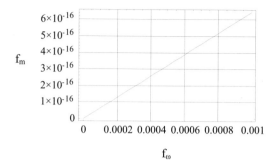

Figure 6.25   Mass fraction in terms of the frequency shift ratio.

Example: Study the relationship between the quantity of added mass and the resonant frequency shift in the case where the deposited mass is localized on a microbridge at $a = 2$ μm. Known are: $\rho = 2200$ kg/m^3, $l = 20$ μm, $w = 2$ μm, $t = 150$ nm, and the resonant shift $\Delta\omega = 1000$ rad/s.

Equation (6.62) is rewritten as

$$\Delta m = 0.0016\frac{1/(1 - f_\omega)^2 - 1}{c_l^4(1 - c_l)^4}m \qquad (6.63)$$

where

$$f_\omega = \frac{\Delta\omega}{\omega_{b,0}} \qquad (6.64)$$

By using the numerical values of this example, the plot of Fig. 6.25 is obtained.

Figure 6.25 indicates that for a fixed position of the attached mass on the microbridge, a higher resonant shift signifies that more mass has attached to the structure (meaning a larger mass fraction f_m).

Layer-mass detection. Only the resonant approach is developed here for layerlike mass deposition detection. Mass which attaches over a portion of a microbridge can be treated by using the model developed for bimorph microbridges. If mass deposition is symmetrically localized with respect to the microbridge midpoint, then Figs. 4.12 and 4.13 can be utilized as well for the analysis that follows. As is the case with layerlike mass deposition on microcantilevers, the similar phenomenon corresponding to microbridges results in alteration of both the stiffness and the mass of the structure, and therefore the change in the resonant frequency results from the interplay between stiffness and mass modifications. By using the half-length model of a dissimilar-length bimorph microbridge, the original stiffness is

$$k_{b,0} = \frac{96(EI_y)_1}{l^3} \tag{6.65}$$

The stiffness after mass deposition becomes

$$k_b = \frac{96(EI_y)_1 (EI_y)_e \, [\, l_p(EI_y)_1 + (l - l_p)(EI_y)_e \,]}{(EI_y)_1^2 l_p^4 + 2l_p(l - l_p)(2l^2 - ll_p + l_p^2)(EI_y)_1 (EI_y)_e + (l - l_p)^4(EI_y)_e^2} \tag{6.66}$$

The effective mass of the original half-length microbridge was given in Eqs. (4.54) as

$$m_{b,0} = \frac{13}{70} m_1 \tag{6.67}$$

where m_1 is the total mass of the substrate. The modified mass is expressed based on Eq. (4.52) as

$$m_b = \frac{13}{70}(m_1 + f_p \Delta m) \tag{6.68}$$

with f_p being defined in Eq. (4.53).

The bending resonant frequency ratio can now be formulated by using Eqs. (6.65) through (6.68) in the form:

$$\frac{\omega_{b,0}}{\omega_b} = 0.277 \sqrt{\frac{m_1 + f_p \Delta m}{m_1} \times \frac{\begin{array}{c}(EI_y)_1^2 l_p^4 + 2l_p(l - l_p)(2l^2 \\ -ll_p + l_p^2)(EI_y)_1(EI_y)_e \\ +(l - l_p)^4(EI_y)_e^2\end{array}}{l^3(EI_y)_e[\, l_p(EI_y)_1 + (l - l_p)(EI_y)_e]}} \tag{6.69}$$

By using the nondimensional parameters

$$f_m = \frac{\Delta m}{m_1} \quad c_t = \frac{t_p}{t_1} \quad c_l = \frac{l_p}{l} \quad c_E = \frac{E_p}{E_1} \quad c_\rho = \frac{\rho_p}{\rho_1} \tag{6.70}$$

Eq. (6.69) becomes

$$\frac{\omega_{b,0}}{\omega_b} = 0.784\sqrt{1 + f_p f_m}\left[(1 - c_l)^3 + \frac{c_l^3(1 + c_E c_t)}{1 + c_E c_t\{4 + c_l[6 + c_t(4 + c_E c_t)]\}} + A\right] \tag{6.71}$$

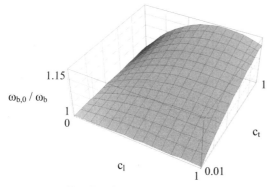

Figure 6.26   Bending frequency ratio in terms of length and thickness ratios.

where
$$A = \frac{3c_l(1 - c_l)(1 + c_E c_t)}{\begin{aligned}&1 + c_E c_t\{4 + c_t[6 + c_t(4 + c_E c_t)]\}\\&- c_l[3 + c_t(6 + c_t(4 + c_E c_t))]\}\end{aligned}} \tag{6.72}$$

Example: Analyze the shift in the bending resonant frequency of a rectangular cross-section microbridge in terms of the nondimensional parameters c_l, c_t, c_E, and c_ρ. Consider the following parameter ranges: $c_l \rightarrow$ [0, 1], $c_t \rightarrow$ [0.01, 1], $c_E \rightarrow$ [0.1, 1], and $c_\rho \rightarrow$ [0.5, 5].

By taking c_l and c_t as variables, also when $c_E = 0.2$ and $c_\rho = 0.5$, the plot of Fig. 6.26 is obtained based on Eq. (6.71). The bending resonant frequency ratio reaches a maximum approximately at the midpoint of the microbridge ($c_l = 0.5$) for every value of c_t, and increases quasi-linearly with the thickness ratio, as Fig. 6.26 shows.

Increasing Young's modulus of the patch relative to the modulus of the substrate results in a slight reduction in the bending resonant frequency ratio because the net effect is a decrease in the overall structural stiffness. Shown in Fig. 6.27 is also the effect of increasing the density ratio, which results in an increase of the bending resonant frequency ratio. This becomes more obvious if the factor in Eq. (6.71) is made explicit in terms of c_ρ, namely,

$$\sqrt{1 + f_p f_m} = \sqrt{1 + f_p c_l c_t c_\rho} \tag{6.73}$$

which indicates the above-mentioned relationship between the resonant frequency ratio and the density ratio.

6.4.2 Variable-cross-section microbridges

The resonant approach is analyzed only for pointlike mass deposition in this subsection. The deposition of pointlike masses on microbridges of variable-cross-section can be studied in a similar manner to that developed for variable-cross-section microcantilevers. Equations (6.45)

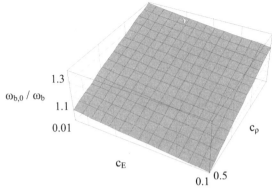

Figure 6.27   Bending frequency ratio in terms of Young's modulus and density ratios.

through (6.47), which define the efficient mass after deposition, the altered resonant frequency, and the quantity of deposited mass, respectively, are still valid, such that the resonant frequency ratio can be expressed as

$$\frac{\omega_{b,0}}{\omega_b} = \sqrt{1 + 256 c_l^4 (1 - c_l)^4 f_m} \tag{6.74}$$

where c_l is the nondimensional length parameter and f_m is the mass fraction, both defined in Eqs. (6.24). Figure 6.28 is the plot of the frequency ratio of Eq. (6.74). The maximum value of this ratio is identical to that corresponding to a variable-cross-section microcantilever, as shown in Fig. 6.28. The profile of the surface plotted in Fig. 6.28 shows that there is a maximum of the frequency ratio at $c_l = l/2$, which indicates that for a given variable-cross-section microbridge, the modified resonant frequency (through the addition of mass) is minimum when the mass attaches at the midpoint. Figure 6.28 also illustrates that the frequency ratio increases with larger mass fractions because more added mass reduces the altered resonant frequency of the microbridge.

The quantity of deposited mass can be calculated in terms of the frequency shift ratio f_ω as

$$\Delta m = \frac{1/(1 - f_\omega)^2 - 1}{256 c_l^4 (1 - c_l)^4} m_{b,0} \tag{6.75}$$

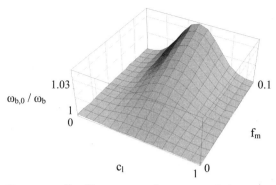

$\omega_{b,0} / \omega_b$

1.03

1

0

c_l

1

0

0.1

f_m

0

Figure 6.28   Bending resonant frequency ratio for a variable-cross-section microbridge in terms of the nondimensional length parameter and mass fraction.

Conversely, the absolute frequency shift can be determined in the form:

$$\Delta\omega_b = \left[1 - \frac{1}{\sqrt{1 + 256c_l^4(1 - c_l)^4 f_m}} \right]\omega_{b,0} \qquad (6.76)$$

Example: Study the quantity of deposited mass which can be detected resonantly by a paddle microbridge with step variable thickness, such as the one pictured in Fig. 4.22, given the defining geometry and material parameters: $l_1 = l_2 = 100$ µm, $w = 20$ µm, $t_1 = 0.2$ µm, $t_2 = 0.5$ µm, $\rho = 2200$ kg/m³, and $E = 160$ GPa.

By using the numerical data of this example in conjunction with Eq. (4.134) giving the stiffness and Eq. (4.136) giving the effective mass, both associated with the midpoint of a paddle microbridge with step variable thickness, it is found that the effective mass of the original structure is $m = 2.216 \times 10^{-12}$ kg and the bending resonant frequency is $\omega_{b,0} = 1.377$ MHz. By using Eq. (6.75) which yields the deposited mass in terms of structural properties and the nondimensional properties, the plot of Fig. 6.29 has been drawn showing the variation of the added mass with the parameters f_ω and c_l. It can be seen that this microbridge design is capable of detecting masses on the order of 10^{-14} kg (tens of femtograms) when the frequency shift ratio (the absolute frequency shift to the original resonant frequency ratio) reaches values in the vicinity of 0.0001. It has been assumed that the mass can attach anywhere on the thickest middle portion.

6.5   Mass Detection by Means of Partially Compliant, Partial-Inertia Microdevices

There are microcantilever- and microbridge-based designs where certain structural segments can be considered rigid, generally because their dimensions (particularly thickness) are larger than those of other

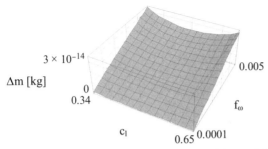

Figure 6.29   Deposited mass in terms of position and frequency shift ratio.

segments which are clearly compliant. Another simplifying assumption that is often made with such microdevices is that mass/inertia comes from only the more massive components (the ones that are also considered rigid) while the inertia fractions of the compliant members are neglected. While this approach, which is based on the two above-mentioned simplifying assumptions, substantially reduces the calculation volume and effort and is in the majority of situations of sufficient accuracy, there might be designs, especially of nanoresonators that are built as thin films where the dissociation rigid-compliant and with-without inertia is more nuanced. In such cases, by not considering the compliance and inertia properties of all components, errors might be generated that could play an important role in the accuracy of predictions.

Two examples are analyzed in this subsection, namely, a paddle microcantilever and a paddle microbridge, by comparing their mass detection performance in the resonant regime when a segment is considered rigid (partially compliant model) and only inertia pertains to it, as compared to the situation where all segments are considered compliant (fully compliant model) and inertia contributions for all segments are accounted for (full-inertia model).

6.5.1 Paddle microcantilevers

The paddle microcantilever of Fig. 6.30 is used to compare the mass detection by means of two models. One is the fully compliant (where the stiffnesses of both segments are taken into account), full-inertia (inertia contributions from both segments are lumped at the free end) model, and the other one is a simplified model which considers only the stiffness of the root segment (the thinner one) and the mass of the wider one as being applied at the end of the root segment.

Obviously, the simplified model is not capable of capturing the location of the deposited mass Δm on the paddle, and it is interesting

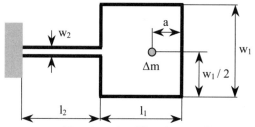

Figure 6.30   Top view of paddle microcantilever with pointlike attached mass.

to check the divergence/convergence of results produced by the two models. Two amounts are monitored: the resonant frequency ratio given in Eq. (6.46) and the quantity of deposited mass which can be calculated by means of Eq. (6.47).

The resonant frequency depends on the lumped-parameter mass of the microcantilever and the quantity of deposited mass. While in a previous example these amounts were determined for a fully compliant, full-inertia model of the paddle microcantilever, the resonant frequency ratio of the simplified model is

$$r\omega_s = \frac{\omega_{b,0}}{\omega_b} = \sqrt{1 + \frac{\Delta m_s}{\rho \omega_1 t l_1}} \tag{6.77}$$

The resonant frequency ratio of the fully compliant, full-inertia paddle microcantilever is

$$r\omega_f = \sqrt{1 + \left[1 - \frac{1.5a}{l_1 + l_2} + \frac{0.5a^3}{(l_1 + l_2)^3}\right]^2 f_m} \tag{6.78}$$

The plot of Fig. 6.31 shows the variation of the following ratio:

$$r_\omega = \frac{r\omega_s}{r\omega_f} \tag{6.79}$$

The variables that appear in Fig. 6.31 are

$$c_l = \frac{l_2}{l_1} \qquad c_w = \frac{w_2}{w_1} \tag{6.80}$$

and the subscripts f and s mean full and simplified, respectively, with reference to the two models. The bending resonant frequency ratios produced by the two models are quite close, as Fig. 6.31 shows. This comparison can, however, be misleading, because it actually analyzes

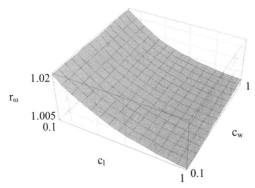

Figure 6.31   Simplified versus full model of paddle microcantilever: comparison between the resonant frequency ratios.

the ratio of two frequency ratios and therefore gives just a relative measure of the differences between the two models.

The other monitored parameter, as mentioned previously, is the quantity of deposited mass, which is provided in generic form by Eq. (6.47). In the case of the simplified model, the quantity of deposited mass is expressed as

$$\Delta m_s = \left[\frac{1}{(1 - f_\omega)^2} - 1 \right] m_s \qquad (6.81)$$

The mass that attaches to the paddle microcantilever according to the fully compliant, full-inertia model is

$$\Delta m_f = \frac{1/(1 - f_\omega)^2 - 1}{[1 - 1.5a/(l_1 + l_2) + 0.5a^3/(l_1 + l_2)^3]^2} m_f \qquad (6.82)$$

where m_f is the effective mass of the paddle microcantilever calculated according to Eq. (3.32). The following ratio is defined:

$$r_m = \frac{m_f}{m_s} \qquad (6.83)$$

By considering the numerical values $l_1 = 200$ μm, $t = 1$ μm, $E = 160$ GPa, $\rho = 2200$ kg/m^3, as well as a value of 0.001 for $\Delta\omega$ (the frequency shift), the mass ratio of Eq. (6.83) is plotted in Fig. 6.32. In both Figs. 6.31 and 6.32, $a = l_1 / 2$.

As Fig. 6.32 indicates, the differences between the two models concerning the prediction of the deposited mass quantity are quite substantial for the parameter ranges considered. For smaller values of

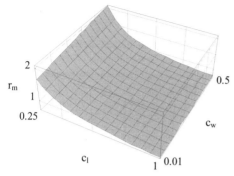

Figure 6.32 Full versus simplified model of paddle microcantilever: comparison between the quantities of detected mass.

the length parameter c_l (the ratio of the thin segment length to the length of the thick segment), the mass predicted by the full model can be twice the mass predicted by the simplified model. However, for designs where the lengths of the two segments are approximately equal and for cases where $l_2 > l_1$ (this situation is not captured in Fig. 6.3), the mass predictions by the two models are similar.

6.5.2 Paddle microbridges

The paddle microbridge which is shown in Fig. 6.33 is analyzed now in terms of mass detection by using an approach similar to that applied to the paddle microcantilever. This time, however, the capacity of detecting mass attachment through monitoring of the torsional resonant frequency is analyzed in addition to bending.

Bending. In bending, the reasoning and approach follow the path described in the subsection treating the paddle microcantilever. The resonant frequency ratio is determined by using a simplified model first whereby the stiffness is produced by only the side (thinner) legs,

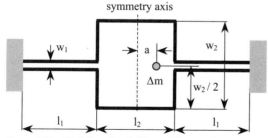

Figure 6.33 Paddle microbridge with mass localized on the longitudinal axis.

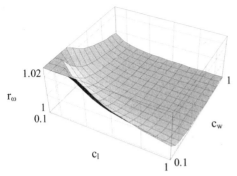

Figure 6.34 Simplified versus full model of paddle microbridge: comparison of the resonant frequency ratios through bending.

whereas the inertia (mass) contribution comes from only the middle segment. A full-compliance, full-inertia model is also analyzed which takes stiffness and inertia contributions from all segments. Both models consider the half-length paddle microbridge.

The simplified model of a half-length paddle microbridge was presented earlier in this chapter where the lumped-parameter bending-related stiffness, mass, and resonant frequency were derived. The bending frequency ratio can therefore be expressed as

$$r\omega_s = \frac{\omega_{b,0}}{\omega_b} = \sqrt{1 + \frac{2\Delta m}{\rho w_2 t l_2}} \tag{6.84}$$

Similarly, the stiffness, mass, and resonant frequency of the fully compliant, full-inertia model are given in Eqs. (4.123), (4.126), and (4.127), respectively. The bending frequency ratio corresponding to this model is

$$r\omega_f = \sqrt{1 + \frac{(l_1 + l_2/2 - 2a)^4(l_1 + l_2/2 + 4a)^2 f_m}{(l_1 + l_2/2)^6}} \tag{6.85}$$

By using Eqs. (6.84) and (6.85), the ratio defined in Eq. (6.79) is formulated and plotted in Fig. 6.34. As was the case with the similar simulation performed for a paddle microcantilever, the ratio of Eq. (6.79) for the paddle microbridge is almost 1, which indicates that the two models produce almost identical bending frequency ratio results. But this is not a clear indication of how the two models do behave with respect to predicting absolute values, such as the detected mass.

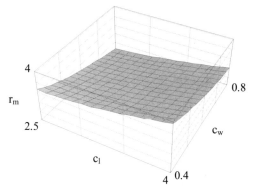

Figure 6.35 Full versus simplified model of paddle microbridge: comparison of the deposited mass quantities through bending.

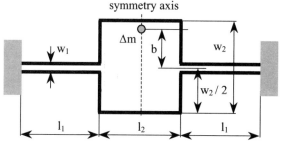

Figure 6.36   Paddle microbridge with eccentrically localized mass.

The mass prediction by the simplified model is given in Eq. (6.81) whereas the mass equation corresponding to the fully compliant, full-inertia, half-length paddle microbridge is

$$\Delta m_f = \frac{1/(1-f_\omega)^2 - 1}{(l_1 + l_2/2 - 2a)^4(l_1 + l_2/2 + 4a)^2/(l_1 + l_2/2)^6} m_f \qquad (6.86)$$

where m_f is the mass given in Eq. (4.126). The mass ratio of Eq. (6.83) is plotted in Fig. 6.35 for the particular value $a = 0$ (which means that the deposited mass is localized at the midpoint of the paddle microbridge) as well as for $l_1 = 200$ μm, $t = 1$ μm, $E = 160$ GPa, $\rho = 2200$ kg/m³, and $\Delta\omega = 0.0001$.

Torsion.   The attached particle can be localized as shown in Fig. 6.36, and in this case the torsional resonance of the paddle microbridge can be utilized to analyze the mass addition phenomenon.

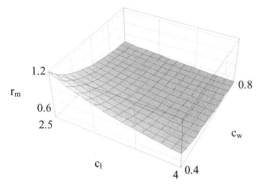

Figure 6.37 Full versus simplified model of paddle microbridge: comparison of the deposited mass quantities through torsion.

A relationship between the original and altered torsional resonant frequencies can be written which is similar to the one applying to the bending resonant frequencies, namely,

$$\frac{\omega_{t,0}}{\omega_t} = \sqrt{1 + f_J} \qquad (6.87)$$

where the inertia fraction is defined as

$$f_J = \frac{\Delta m b^2}{2J_t} \qquad (6.88)$$

J_t is the lumped-parameter torsional moment of inertia of the half-length microbridge, and therefore the factor of 2 in the denominator indicates that only one-half of the mass is taken into consideration. The mass which is sensed by the microbridge can be expressed as

$$\Delta m = \frac{2J_t}{b^2}\left[\frac{1}{(1-f_\omega)^2} - 1\right] \qquad (6.89)$$

By comparing the simplified model which takes into account the stiffness of only the end segments (the middle segment is considered rigid) and the inertia of the middle segment, as mentioned earlier, to the fully compliant, full-inertia paddle microbridge model, the mass ratio introduced in Eq. (6.83) can be plotted as shown in Fig. 6.37.

The same numerical values utilized for the similar plot corresponding to a paddle microcantilever (Fig. 6.35) have been used for the simulation shown in Fig. 6.36. As can be seen, differences appear again between the detected mass according to the two models' predictions, such that for relatively thinner and narrower end segments, the mass prediction by the full model can be 1.2 times larger than the result yielded by the simplified model. For thicker and wider end segments, the relationship reverses and the mass predicted by the full model is approximately 0.6 times the mass obtained through the simplified model.

References

1. R. Raiteri, M. Grattarola, H.-J. Butt, and P. Skladal, Micromechanical cantilever-based biosensors, *Sensors and Actuators B*, **79**, 2001, pp. 115–126.

2. C. L. Britton, R. L. Jones, P. I. Oden, Z. Hu, R. J. Warmack, S. F. Smith, W. L. Bryan, and J. M. Rochelle, Multiple-input microcantilever sensors, *Ultramicroscopy*, **82**, 2000, pp. 17–21.

3. D. R. Baselt, B. Fruhberger, E. Klaassen, S. Cemalovic, C. L. Britton, S. V. Patel, T. E. Mlsna, D. McCorkle, and B. Warmack, Design and performance of microcantilever-based hydrogen sensor, *Sensors and Actuators B*, **88**, 2003, pp. 120–131.

4. I. Dufour and L. Fadel, Resonant microcantilever type chemical sensors: analytical modeling in view of optimization, *Sensors and Actuators B*, **91**, 2003, pp. 353–361.

5. E. Garcia, N. Lobontiu, Y. Nam, B. Ilic, and T. Reissman, Shape optimization of microcantilevers for mass variation detection and AFM applications, *SPIE04 International Conference*, vol. 5390, 2004, pp. 421–427.

6. K. B. Brown, W. Allegretto, F. E. Vermeulen, R. P. W. Lawson, and A. M. Robinson, Cantilever-in-cantilever micromachined pressure sensors fabricated in CMOS technology, *1999 IEEE Canadian Conference on Electrical and Computer Engineering*, 1999, pp. 1686–1691.

7. H. Kawakatsu, H. Toshiyoshi, D. Saya, and H. Fujita, A silicon based nanometric oscillator for scanning probe microscopy operating in the 100 MHz range, *Japanese Journal of Applied Physics*, **38**, 1999, pp. 3962–3965.

8. B. Rogers, L. Manning, M. Jones, T. Sulchek, K. Muray, B. Beneschott, J. D. Adams, Z. Hu, T. Thundat, H. Cavazos, and S. C. Minne, Mercury vapor detection with a self-sensing, resonating piezoelectric cantilever, *Review of Scientific Instruments*, **74**(11), 2003, pp. 4899–4901.

9. L. Pinnaduwage, V. Boiadjiev, J. E. Hawk, and T. Thundat, Sensitive detection of plastic explosives with self-assembled monolayer-coated microcantilevers, *Applied Physics Letters*, **83**(7), 2003, pp. 1471–1473.

10. B. Ilic, D. Czaplewski, H. G. Craighead, P. Neuzil, C. Campagnolo, and C. Batt, Mechanical resonant immunospecific biological detector, *Applied Physics Letters*, **77**(3), 2000, pp. 449–451.

11. B. Ilic, D. Czaplewski, M. Zalatudinov, and H. G. Craighead, Single cell detection with micromechanical oscillators, *Journal of Vacuum Science Technology*, **19**(6), 2001, pp. 2825–2828.

12. L. Sekaric, J. M. Parpia, H. G. Craighead, T. Feygelson, B. H. Houston, and J. E. Butler, Nanomechanical resonant structures in nanocrystalline diamond, *Applied Physics Letters*, **81**(23), 2002, pp. 4455–4456.

13. L. Sekaric, M. Zalatudinov, R. B. Bhiladvala, A. T. Zehnder, J. M. Parpia, and H. G. Craighead, Operation of nanomechanical resonant structures in air, *Applied Physics Letters*, **81**(14), 2002, pp. 2641–2643.

14. N. V. Lavrik and P. G. Datskos, Femtogram mass detection using photothermally actuated nanomechanical resonators, *Applied Physics Letters*, **82**(16), 2003, pp. 2697–2699.

15. B. Ilic, Y. Yang, and H. G. Craighead, Virus detection using micro-electromechanical devices, *Applied Physics Letters*, **85**(13), 2004, pp. 2604–2606.

16. B. Ilic, H. G. Craighead, S. Krylov, W. Senaratne, C. Ober, and P. Neuzil, Attogram detection using nanoelectromechanical devices, *Journal of Applied Physics*, **95**(7), 2004, pp. 3694–3703.

17. S. Evoy, A. Olkhovets, L. Sekaric, J. M. Parpia, H. G. Craighead, and D. W. Carr, Time-dependent internal friction in silicon nanoelectromechanical systems, *Applied Physics Letters*, **77**(15), 2000, pp. 2397–2399.

18. K. Y. Yasumara, T. D. Stowe, E. M. Chow, T. Pfafman, T. W. Kenny, B. C. Stipe, and D. Rugar, Quality factors in micron- and submicron-thick cantilevers, *Journal of Microelectromechanical Systems*, **9**(1), 2000, pp. 117–125.

19. J. Yang, T. Ono, and M. Esashi, Mechanical behavior of ultrathin microcantilever, *Sensors and Actuators*, **82**, 2000, pp. 102–106.

20. J. Tamayo, A. D. L. Humphris, A. M. Malloy, and M. J. Miles, Chemical sensors and biosensors in liquid environment based on microcantilevers with amplified quality factor, *Ultramicroscopy*, **86**, 2001, pp. 167–173.

21. N. Lobontiu and E. Garcia, *Mechanics of Microelectromechanical Systems*, Kluwer Academic Press, New York, 2004.

Index

Accelerometers, resonant, 284–289
AFM (atomic force microscopy), 2
Amplitude ratio (at resonance), 6–7
Arbitrarily translated reference frames,
    compliances in, 110–111
Atomic force microscopy (AFM), 2
Axial distribution function, 56
Axial load, long cantilever with, 94–98
    Rayleigh-Ritz procedure, 96–98
    Rayleigh's procedure, 95–96
Axial-related stiffness, 49–50, 53, 54
Axial vibrations:
    for constant-cross-section members, 59–60
    and lumped-parameter inertia, 56–57

Bandwidth, 8
Beam-type microresonators, 227–258
    examples of, 245–258
    microbridge resonant frequency models,
        237–245
    microcantilever resonant frequency
        models, 228–237
        paddle microcantilever of constant
            thickness, 229–233
        paddle microcantilever of constant
            width, 233–237
Bending-related stiffnesses, 47–48, 51–55
Bending resonant frequency:
    of constant-cross-section microbridges,
        169–175
        long microbridges, 170–172
        short microbridges, 172–175
    of serially compounded microbridges
        three-segment microbridges, 196–199
        two-segment microbridges, 192–194
Bending vibrations:
    for constant-cross-section members,
        64–71
        long microcantilevers, 64–68
        short microcantilevers, 68–71
    and lumped-parameter inertia, 58
Bimorphs, 151, 158
Bridges, micro- (see Microbridges)

Cantilevers, micro- (see Microcantilevers)
Castigliano's first theorem, 48–49
Castigliano's second (displacement)
    theorem, 52–53, 160
Circular cross-section microbars, 61–63

Circularly filleted microcantilevers:
    long, 126–130
    thin, 83–87
Circularly filleted microhinges:
    corner, circular, 139–142
    right circular, 134–137
Circularly notched microcantilevers, 142–145
Circular rings, 98–101
Clamping losses, 19, 20
Classical tuning fork, 2
Comb-finger transduction, 264
Complex impedance (Z), 39–41
Compliance (flexibility) method, 24–26,
    52–55
Compliance matrix, 53–54
Compliance transforms:
    for microhinges/microcantilevers, 107–111
        in arbitrarily translated reference
            frames, 110–111
        in opposite-end reference frames,
            108–110
    for variable-cross-section microbridges,
        185–186
Compliant segments, micromembers formed
    of two, 111–145
    circularly notched microcantilevers,
        142–145
    filleted microcantilevers, 126–134
    filleted microhinges, 134–142
        circular corner-, 139–142
        right circular, 134–137
    paddle microcantilevers, 115–126
Constant-cross-section members, 58–71
    axial vibrations, 59–60
    bending vibrations, 64–71
        long microcantilevers, 64–68
        short microcantilevers, 68–71
    torsional vibrations, 61–63
Constant-cross-section microbridges,
    171–180
    bending resonant frequency of, 169–175
    mass detection with, 320–326
        layer-mass detection, 323–326
        point-mass detection, 320–323
    torsional resonant frequency of, 175–178
Constant-cross-section microcantilevers,
    mass detection with, 304–315
    layer-mass detection, 308–315
    point-mass detection, 304–308
Corner-filleted microcantilevers, 82–91
    circularly filleted microcantilever, 83–87

337

ABOUT THE AUTHOR

Nicholae Lobontiu, Ph.D., is currently a professor at the Technical University of Cluj-Napoca, Romania. He is also a research associate and instructor at the Sibley School of Mechanical and Aerospace Engineering, Cornell University. He has done extensive work in the areas of MEMS, monolithic hinges, and compliant mechanisms. Dr. Lobontiu is the author of two prior books on hinge-based compliant mechanisms and the mechanics of MEMS and has published numerous journal articles and has given several presentations on the subject of nano/micro system modeling and design.